Worked Examples in Quantity Surveying Measurement

Peter Goodacre

and

William Crosbie-Hill

London   New York

E. & F. N. SPON

First published 1982
by E. & F.N. Spon Ltd
11 New Fetter Lane, London EC4P 4EE

Published in the USA by
E. & F.N. Spon
733 Third Avenue, New York NY 10017

© *1982 Peter Goodacre and William Crosbie-Hill*

Printed in Great Britain by
J. W. Arrowsmith Ltd, Bristol

ISBN 0 419 12340 7

British Library Cataloguing in Publication Data

Goodacre, Peter E
    Worked examples in quantity surveying
    measurement.

    1. Building — Estimates
    I. Title   II. Crosbie-Hill, William
    692'.5   TH435

    ISBN 0-419-12340-7

Library of Congress Cataloging in Publication Data

Goodacre, Peter E
    Worked examples in quantity surveying
    measurement.

    Includes index.

    1. Building — Estimates — Problems, exercises, etc.
    I. Crosbie-Hill, William.   II. Title.
    TH437.G64 1982   692'.5   82-9508
    ISBN 0-419-12340-7   AACR2

Worked Examples in Quantity Surveying Measurement

Contents*

Items in parentheses refer to drawing numbers or descriptions

To A. L. M.

The Authors

Peter Goodacre, RD, MSc, FRICS, is a senior lecturer in quantity surveying in the Department of Construction Management at the University of Reading. Having worked for a firm of Chartered Quantity Surveyors prior to taking up his appointment at the University, he remains a consultant to both private practice and industry. He has strong links with the RICS, serving for four years on the Quantity Surveyors Divisional Council as well as continuing to work on other various Institution activities. Educated at Ewell Technical College, the College of Estate Management and Loughborough University, his current research interests lie in the early stages of building design. The author of several other books, he has lectured widely to professional audiences, most recently on the Standard Form of Contract and the Sixth Edition of the Standard Method of Measurement.

William Crosbie-Hill, ARICS, is Principal Quantity Surveyor for the London Borough of Islington. Previously Chief Quantity Surveyor at Brent and Senior Quantity Surveyor with the National Building Agency his work has been mainly concerned with the redevelopment of deteriorating urban areas. Educated at the College of Estate Management he has maintained an interest in the training of quantity surveyors and has been a visiting lecturer in quantity surveying in the Department of Construction Management at the University of Reading. In the last few years he has been an active member of the Society of Chief Quantity Surveyors in Local Government.

Introduction

It is intended that these worked examples serve two purposes. The first purpose, and that for which they were originally devised, is to provide a detailed approach to some realistic measurement problems. To this end a number of the drawings have been taken from real projects. The student is then faced with the task of identifying and selecting the relevant information from the drawings and where appropriate raising queries with the architect or engineer. The examples are also longer than those normally encountered in quantity surveying textbooks. This reflects the approach of the authors to get students to see the whole problem rather than spend a disproportionate time, never allowable in practice, over an item of detail.

The second purpose that it is hoped these examples will serve is a guide to practitioners. We do not say that there is no better solution to the various problems than ours. However what we do say is that they represent good practice, not only in the way individual items have been measured but also in the way the work is set out. Much of the work involved in measurement is of a routine nature and to some extent its importance in education has been denigrated in recent years. However, we believe routine work such as this has a place in higher education. Much of the work of other professions, the law, medicine, engineering, accountancy, etc., is of a routine nature, but it is with well established basic methods that the standards of those professions are maintained.

The eight examples are divided into three sections. The first section contains two examples applied to the structure of complete small buildings, giving an overview of a large number of SMM items. The second section contains three examples illustrating more complex structural elements of a building. The final section covers the measurement of services installations which is thought to be an area insufficiently covered by the standard text books on measurement.

As an aid to the student a taking-off list precedes the example where there are several different sections of work to be measured. While this is not always necessary for the practitioner, much of the measurement we have seen is poorly organized and would have benefited from such a list.

Where items have no quantity they have not been written across the dimension columns but instead the word 'Item' is put in the dimension column. Some offices adopt this practice and it is common in computer and cut and shuffle systems. The main purpose in this book for adopting this convention has been one of making students aware of the distinction between headings and descriptions.

References to the appropriate clauses in SMM6 are given after each item. Obviously these would not be included in a real problem. However we have been concerned at the lack of understanding of SMM6 due simply to not having read it.

As it is essential that this book is used in conjunction with SMM6, the index (p.217) follows the SMM6 layout rather than conforming to the usual alphabetical style.

For clarity we have not used abbreviations in descriptions although they would be used in practice. We have used traditional dimension paper, this being the simplest way to show the logical stages of measurement. Once the logic is understood then variations in either paper format and/or style of descriptions, such as 'cut and shuffle' paper, 'Fletcher and Moore Standard Phraseology', computerized systems, etc., can be applied.

Finally, we are grateful to the publishers for agreeing to take on a difficult production task and, in particular, we would like to thank Philip Read for his encouragement and patience. Considerable thanks are due to Alfred Head, Esq., RIBA, Borough Architect, London Borough of Islington, for the use of drawings on which most of the examples are based. Some of the other drawings originated from John Kelly and Roger Flanagan to whom thanks are also due for permission to use them in this work. James Smith of Robinson and Roods, Chartered Quantity Surveyors, gave invaluable advice on the adequacy of individual descriptions. The work was typeset and produced by the College of Estate Management; a difficult task undertaken by a large, enthusiastic team led by John Armstrong and Leonard Moseley. John Jewell redrew some of the drawings.

Whilst there are no deliberate errors, there will inevitably be some errors in a work of this nature. The authors will be pleased to hear from any readers spotting any errors or omissions.

January 1982

Peter Goodacre
William Crosbie-Hill

Notes on method of measurement

Where by its context in the description a dimension is in millimetres, the abbreviation mm is usually omitted. For example, '20 Asphalt in two coats as roof covering' is deemed to read '20 mm thick Asphalt . . .'

Brickwork etc., is measured over all door and window openings and adjustments for this over-measurement are made with the measurement of windows and doors.

The word 'ditto' has been used in descriptions to mean the same as the preceding item. The following rules in the use of ditto should be applied:

1. Ditto should never be used unless its meaning is absolutely clear.
2. Ditto should not be used for single words.
3. Preceding items can be qualified by using Ditto *and* or Ditto *but* when the addition or alteration is obvious by its context.

Drawings

The drawings on which the worked examples are based form an integral part of this book, and A4 reductions have been included in the text to permit quick reference to key items. However, to allow students and lecturers to study the full-sized drawings, these have been published separately as a boxed set and are available from the publishers.

Section One
Domestic Construction

EXAMPLE ONE

Single storey structure 1

Project: Garage and office
Drawings: Garage and office (JK/1)

Generally

This is a simple structure in which the foundations, superstructure, roof, partitions, finishings, and windows are measured. The drawings would be provided to comply with clause B.3.1. Therefore there is no need to provide general descriptions of the work to comply with G.1, L.1, M.1, P.1, etc.

Column 3

The oversite is measured to the outside edge of the external wall at the door entrance. The additional oversite for the jambs of the doors is taken in Column 5.

Column 5

The additional area of oversite excavation, breaking up and making good of the concrete by the garage doors is taken here.

Column 7

The dimension 5.40 is the inside dimension between the trench excavations and is taken from the cross wall dimension in Column 4. The 2.40 dimension is the net length of the dividing wall between the WCs. The 800 mm is the average width of the thickening.

Columns 8 and 9

There is no requirement in SMM6 to repeat the plant items when the superstructure is measured. However, many practices bill the superstructure under a separate heading (i.e. they do not aggregate quantities of brickwork in the substructure with that in the superstructure). In this case it does then become necessary to measure the plant items twice and this is the approach adopted in this example.

2

Column 9

The first adjustment (less 2/300) to the centre line girth is to deduct the projection of the foundation beyond the door jambs. The second adjustment (plus 4/265) moves the centre line out onto the exposed face of the outer skin. The third adjustment finds the centre line for the outer skin. A similar procedure is adopted for the inner leaf. In the pamphlet Questions and Answers*, the rule in clause G.4.1 is clarified in that when walls are built in facing bricks they are measured on the centre line and not on the face.

Column 10

The facings are taken for a height of 225 mm (150 from ground level up to dpc and one course below ground level).

Column 12

The backfilling to the outside trench is taken to the existing ground level.

Column 13

The sequence of adjustments is:

1. Centre line less thickness of wall to give inside perimeter of wall including projection beyond jamb opening.
2. Less width of projection (4/300) to give centre line of projection.
3. Add pier adjustment (2/2/215).
4. Less cross-wall intersection (2/215).
5. Add width of wall for return at jamb (2/265) as below.

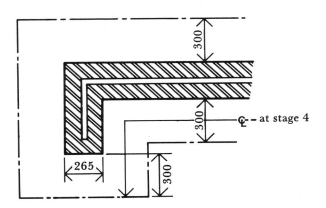

* *Standing Joint Committee for the Standard Method of Measurement (1981), Standard Method of Measurement, 6th edn: Questions and Answers. RICS and NFBTE, London.*

Column 20

The girth for the outer skin is the centre line of the skin. The deducts of 2/300 are from the centre line of the trench which includes a projection beyond the jambs. The deduct for the corners of the inner skin are:

½/55 + ½/105
= 17.5 + 52.5
= 80 mm

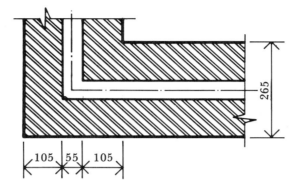

Column 23

The separation of the parapet wall into the heading of non-loadbearing superstructure is in strict accordance with G.3. of the *Practice Manual**. The surveyor should use his discretion when measuring to avoid unnecessary and tedious division of brickwork into load and non-loadbearing classifications.

Column 26

The plinth stretcher to the parapet is measured as an extra over item, using clause G.17.1 as a guide, but not describing it as facework.

Column 27

The location drawings provide the specific information required in L.1. and M.1. There is no restriction on the siting of the pots.

Column 28

The gutter is measured as a full value item, hence the area of the roof is adjusted accordingly. The classification of woodwork as roof decking is given in the Practice Manual.

* *Standing Joint Committee for the Standard Method of Measurement (1978), Practice Manual SMM6. RICS and NFBTE, London.*

Column 30

There is no limiting girth to distinguish between a gutter and a shallow channel (see L.7.3 in the Practice Manual). In this particular example there is an associated skirting and therefore the item has been measured full value (i.e. gutter) and not extra over (i.e. shallow channel).

Column 33

A useful rule of thumb for pipe girths is

        up to 50 mm diameter   = not exceeding 150 mm girth
        up to 100 mm diameter = over 150 mm and not exceeding 300 mm girth

Column 33

For a more detailed commentary on the clauses under which the asbestolux is measured see Column 43.

Column 35

Tapped holes are necessary as nuts cannot be fixed once the beam is put in position.

Column 37

The sockets which are cast in are measured in accordance with clause F.24. There is no need to measure them separately from the padstones. All the steelwork is primed with red oxide paint before fixing. The steel beam for the garage lifting equipment needs a finishing coat as well.

Column 39

There is no need to state the diameter of the bolt as it is the thickness of the timber which is the controlling factor.

Column 40

Although the partition between the office and WCs is full height it is non-loadbearing. If the blockwork is of the hollow block type an item would be measured for filling the partition wall end.

Column 41

The surveyor must use his discretion whether to measure timber stud partitions under clauses N.2.1.b and N.15 or T.20 (dry linings and partitions). In this example, clauses N.2.1.b and N.15 are obviously applicable and one can imagine appropriately obvious situations where T.20 would be used (for example the partitioning to a complete office block). However, if any doubt exists, a note as to the method used should be put in the Bill. If there is any difficulty over the method used and its effect on pricing the estimator can raise this issue before tenders are submitted. The sequence of measuring the studs is:

1. Verticals
2. Top rails
3. Bottom rails
4. Intermediate studs
5. Intermediate studs over door openings.

Column 43

Section T of the Standard Method includes *in situ* finishings and flexible sheet finishings but not rigid finishings. In the pamphlet, *Questions and Answers on the 6th Edition* * it should be noted that plasterboard is not measured within the rules described in clause T.16.1.a. Therefore plasterboard, asbestolux, etc. are presumably measured under clause N.15.

Column 50

It is probably not necessary to have as full a description as this plus the reference to the drawing. An alternative approach is to identify the window clearly enough in the bills so that easy reference can be made to the drawing or catalogue.

Column 55

The vertical closing of the cavity includes the vertical damp proof course. It is not clear whether all work in compartments not exceeding 4.00 m² (e.g. narrow width, angle beads, etc.) should be so described. The procedure adopted here is to measure all the plasterwork under the appropriate heading but to group angle beads together. A note to this effect (or to whatever approach has been adopted) should be included in the Bills.

* *Standing Joint Committee for the Standard Method of Measurement (1981), Standard Method of Measurement, 6th edn: Questions and Answers. RICS and NFBTE, London.*

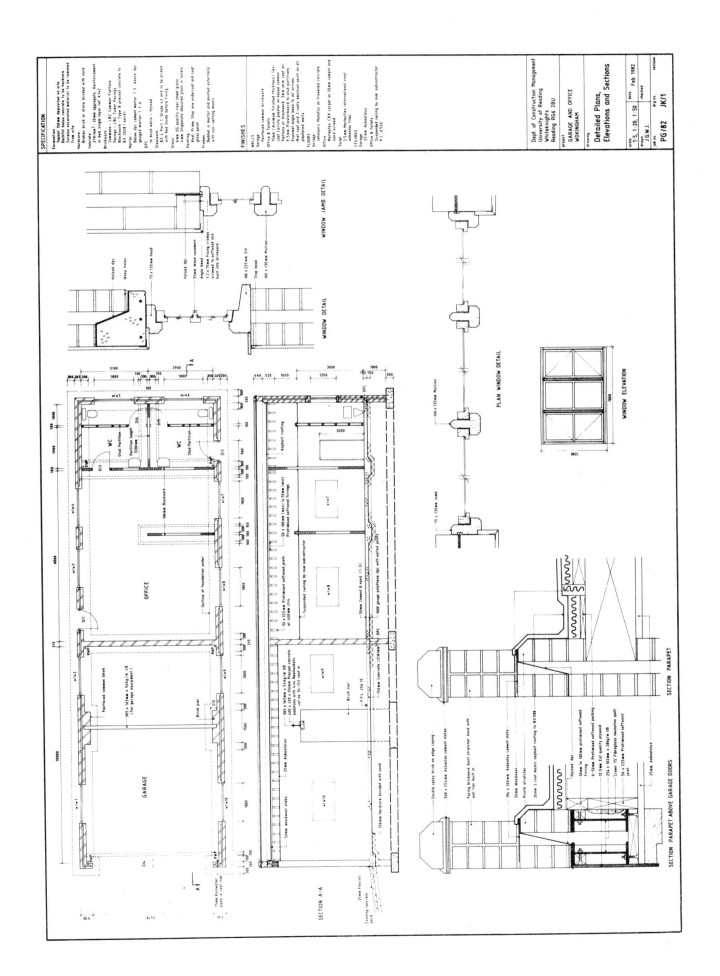

WINDOW JAMB DETAIL

WINDOW DETAIL

SECTION A-A

PLAN WINDOW DETAIL

WINDOW ELEVATION

SECTION PARAPET

SECTION PARAPET ABOVE GARAGE DOORS

GARAGE

OFFICE

WC

WC

Dept of Construction Management
University of Reading
Whiteknights
Reading RG6 2BU

project
GARAGE AND OFFICE
WOKINGHAM

drawing
Detailed Plans,
Elevations and Sections

scale 1:5, 1:20, 1:50   date Feb 1982
drawn J.G.W.J.   checked
job no PG/82   drg no JK/1   revision

SPECIFICATION

Excavation
Topsoil 150mm deposited on site
Backfill below concrete to be hardcore
Surplus excavated material to be removed
from site

Hardcore
Broken brick or stone blinded with sand

Concrete
21N/mm2 - 20mm aggregate Reinforcement
in bed, single layer ref A142

Brickwork
Common - LBC Common Flettons
Facings - LBC Tudor Facings
Blockwork - Type B precast concrete to
BS 2028 (solid)

Mortar
Below dpc cement mortar 1:3 Above dpc
gauged mortar 1:1:6

DPC
In brick walls - Hyload

Brickwork
BS 4 Part 1, Grade 43 and to be primed
with Red Oxide before fixing

Glass
6mm OQ quality clear sheet glass
4mm Stopapyle obscured glass in toilets

Painting
Knot Prime, Stop one undercoat and coat
gloss paint

Frames
Bedded in mortar and painted externally
with non-setting mastic

FINISHES

WALLS
Garage
Fairfaced common brickwork
Office & Toilets
15mm (minimum finished thickness) Two
coat Carlite plaster on keyed common
Flettons or blockwork (3mm one coat on
9.5mm Plasterboard to stud partitions
Expanded angle bead to all arrises
Mist coat and 2 coats emulsion paint on all
plastered walls

FLOORS
Garage
Johnson's Monofix on trowelled concrete
Office
Marleyflex CBX carpet on 50mm cement and
sand screed
Toilet
25mm Marleyflex international vinyl
asbestos tiles

CEILINGS
Garage
25mm Asbestolux
Office & Toilets
Suspended ceiling by non-subcontractor
P.C. £550

GARAGE AND OFFICE

SUBSTRUCTURE

Taking off list

General heading
Ground water
Excavation plant
Surface excavation
Trench excavation
Break up & reinstate yard
Earthwork support
Level and compact
Thickening
Concrete heading
Concrete plant
Concrete foundations
Brickwork Plant
External walls
Internal walls
Damp proof course
Backfilling
Hardcore bed
Concrete bed
Finish to concrete
Expansion joint
Protection

	Item	Three trial holes were dug on the site of the proposed building on 6 May 1979 and ground water was found at an average level of 204·50 . The post-contract water level will be established before commencing on site
		(D.3.1.6)
	Item	Bring to site and remove on completion all plant required for 'Excavation and Earthwork'
		(D.4.1)
		&
		Maintain on site all plant for 'Excavation & Earthwork'
		(D.4.2)

2

Topsoil

100 00	3100		
215	2900		
80 00	265		
100	265		
1900	300		
100	300		
1000	7130		
265			
265			
300			
22 145			

Girth

	10000	
	8000	
	1900	
	1000	
2/100	215	
2/265	200	
	530	
	3100	
	2900	
2/265	530	
	28375	
	2	
	56750	
Less		
4/265	1060	
	55690	
Less		
5470		
2/300 600	4870	
	50 820	

```
22·15
7·13
```

Excavate topsoil
average 150 mm
deep

(D.9)

&

Deposit topsoil on
site average 150 mm
thick

(D. 30/0.36)

```
50·82
0·87
1·00
```

Excavate trench
starting at surface
strip level to
receive foundations
over 0·30 m in
width & maximum
depth not exceeding
1·00 m

(D. 13. 6. b)

&

Remove surplus
excavated material
from site

(D. 29)

Trench

Depth	1000	
	300	
	1300	
Less 2/150	300	
	1000	
Width	265	
2/300	600	
	865	

```
5·40
0·82
1·00
```

```
2/ 2·10
0·22
1·00
```

Cross-wall

	3100	
	2800	
	6000	
Less 2/300	600	
215	5400	
2/300 600		
815	Piers	
	1500	
2/300	600	
	2100	

3

4

GARAGE AND OFFICE

Yard

2/	1·13	
	0·30	
	0·15	

Excavate oversite to
reduce levels

(D. 13. 3)

&

Remove surplus
excavation from
site

(D. 29)

	530
2/300	600
	11 30

2/	1·13	
	0·30	

Extra over reduce
level excavation
for breaking up
existing concrete
bed 150 mm thick

(D. 13, 11)

2/	1·13	
	0·30	
	0·15	

Concrete (1 : 2 : 4 –
20 mm aggregate)
bed 100 – 150 mm
thick in making
good edge of
existing yard

(F. 6. 8)

5

Earthwork support

2/	50·82	
	1·00	

Earthwork support
to trenches
maximum depth
not exceeding
1·00 m & not
exceeding 2·00 m
between opposing
faces

(D. 18)
(x-wall

2/	5·40	
	1·00	

(piers

2/2	0·22	
	1·00	

(ends
of trench

2/	0·87	
	1·00	

2/	0·82	
	1·00	

Ddt

Ditto   (Intersect
of x-wall

Level & compact

50·82	
0·87	

5·40	
0·82	

Level & compact
bottom of
excavation   (x-wall
(D. 40)

2/	2·10	
	0·22	

(piers

Item	

keep excavation
free from surface
water

(D. 25)

6

		Thickening	
		5400	
		1000	
		1900	
		100	
	Less	3000	
	2/300	600	2400
			7800
	Less	5470	700
	2/300	600	900
		4870	2|1600
			800

7.80 0.80 0.15	Excavate trench starting at surface strip level to receive foundations over 0.30 m in width and maximum depth not exceeding 0.25 m (D.13.6.b)
2.90 0.80 0.15	
	& (Entrance
4.87 0.67 0.15	Remove excavated material from site (D.29)

7

Other Concrete Work

(Total volume of insitu Concrete = approximately ___ m³)

(F.3.2)

Item	Bring to site and remove on completion all plant required for 'Concrete Work' (F.2.1) & Maintain on site all plant for 'Concrete Work' (F.2.2)

Foundations

50.82 0.87 0.30 5.40 0.82 0.30 2/ 2.10 0.22 0.30	Plain ordinary Portland cement concrete ($21 N/mm^2$ – 20 mm aggregate) foundations in trenches 150-300 mm thick poured against faces of excavation (F.6.2)
50.82 0.06 0.85	Concrete (1:10 – 10 mm aggregate) filling to cavity not exceeding 100 mm thick (F.6.18)

8

GARAGE AND OFFICE

Brickwork

Item	Bring to site and remove on completion all plant required for 'Brickwork and Blockwork'	(G.2)

&

Maintain on site all plant for 'Brickwork & Blockwork'	(G.2)

Brickwork in Foundations

(G.3.1.a)

External walls

Outer leaf

Centreline girth		50 820
Less ²/300		600
		50 220
⁴/265		1 060
Outer face		51 280
		51 280
Less ⁴/105		420
		50 860

Inner Leaf

Girth		50 820
Less ²/300		600
		50 220
Less ⁴/265		1 060
Inner face		49 160
		49 160
Plus ⁴/105		420
		49 580

9

50·86	Half brick skin of
0·78	hollow wall in
49·58	LBC Flettons in
1·00	Stretcher bond in cement mortar (1:3)
	(G.5.3.c)

Height	1000
Less 150	
75	225
	775

50·22	Form 55 mm wide
1·00	cavity in hollow wall with galvanised twist pattern wall ties to BS 1243 Type 'A' at the rate of 5 per square metre
	(G.9.1)

50·86	Half brick skin of
0·23	hollow wall in 'Tudor' facings in Stretcher bond in cement mortar (1:3) and point with a weathered joint as the work proceeds
	(G.14.9)

²/ 1·00	Projection of attached pier in LBC Flettons in English bond in cement mortar (1:3) 1500 wide × 215 mm depth at projection
	(G.5.4)

10

13

Dimension Column 1 (page 11)

		Internal wall
	6·00	One brick wall in
	1·00	LBC Flettons in
		English bond in
		cement mortar (1:3)
		(G.5.3.a)
		Damp proof
		Courses
2/	50·28	Horizontal 'Hyload'
		pitch polymer damp
		proof course 110 mm
		wide laid with joints
		lapped 150 mm and
		bedded in cement
		mortar (1:3) (no
		allowance made for
		laps)
		(G.37.2)
2/	1·50	Ddt
		Ditto (Piers
2/	0·85	Close 50 mm wide
		cavity with LBC
		Fletton brickwork
		half brick deep
		(G.9.2)
2/	0·15	Ditto but with
		Tudor facings

11

Dimension Column 2 (page 12)

2/	1·50	Horizontal 'Hyload'
	0·33	pitch polymer
		damp proof course
		laid with joints
		lapped 150 mm
		and bedded in
		cement mortar
		(1:3) (no
		allowance made
		for laps)
		(G.37.2)
		(Piers
2/	6·00	Ditto but 215 mm
		wide
		(G.37.2)
		Backfilling
		(a) Outside trench
		External face 56 750
		Less 5 470
		51 280
		Less 2/530 1 060
		50 220
		2/300 600
		50 820
50·82	Ddt
0·30	Remove surplus
0·85	as before
		&
		Add
		Filling to excavations
		in selected
		excavated material
		(D.35)
		1000
		Less 300
		700
		150
		850

12

14

GARAGE AND OFFICE

Backfilling (cont)

(a) Outside trench (cont)

		530
2/300		600
		1130

2/ 1.13
0.30
0.70

Filling to excavations in broken brick or stone

(D.35)

(Under new yard bed

(b) Inside trench

Centre line		50 820
Less 4/265		1060
		49 760
Less 4/300		1200
		48 560
Add		
2/2/215		860
		49 420
Less		
2/215		430
		48 990
Add		
2/265		530
		49 520

Note — no adjustment made for thickening of slab where this intersects with hardcore backfilling

13

49.52
0.30
0.70

Filling to excavations in broken brick or stone

(D.35)

(x-wall

2/ 5.40
0.30
0.70

Hardcore bed

		8000
		1900
		1000
2/100		200
		11100

10.00
6.00
11.10
6.00
5.47
0.27

Broken brick or stone filling to make up levels average 150 thick well compacted

(D.36/D.37)

(sinkings

2/ 11.90
0.10

&

5.47
0.10

Blind filling with sand to receive concrete

(D.43)

		6000
	1900	
	100	
	1000	
	3000	
Less	300	2700
	2900	
	300	3200
		11900

2/ 1.50
0.22

Ddt

Last two items

(Piers

14

15

Hardcore bed (cont)

6.00 — Hand packing hardcore to form battering face 150 mm high (D.38)

11.90 — Ditto to form sinking average 800 wide × 150 mm deep (D.39)

Concrete bed

Reinforced ordinary Portland cement concrete (21 N/mm² – 20 mm aggregate) bed 100 – 150 mm thick

```
        10.00
         6.00

        11.10
         6.00

         5.47
         0.27                   (thickening
                          Cube × 0.15 = _____
        11.90
         0.80                   (Entrance
                                 to garage
         5.47
         0.60                    (Ditto

  2/     0.27                    (F.6.8)
         0.60
```

```
  2/     1.50     Ddt
         0.22        Ditto
                     Cube × 0.15 = _____

                                (Piers
```

```
        11.10
         6.00
  2/    11.90
         0.10
```
Horizontal 1000 gauge polythene membrane with welted seam joints laid on prepared bed (measured net – no allowance made for laps) (G.37.2)

```
        10.00
         6.00
        11.10
         6.00
         5.47
         0.27
```
Mild steel fabric reinforcement to BS 4489 Reference A 142 weighing 2.22 kg/m² laid in ground slabs including 150 mm side and end laps (F.12.3)

```
  2/     1.50     Ddt
         0.22        Ditto          (Piers
```

5.47 — Ditto single strip 450 mm wide (F.12.4)

11.90 — Ditto single strip 800 mm wide (F.12.4)

15   16

GARAGE AND OFFICE

<u>Finish</u>

<u>Protection</u>

10·00 6·00	Trowel finish to surface of unset concrete bed
5·47 0·27	(F. 9. 3)

Item

Protect the whole of the work in the 'Substructure' section

(D.45 etc)

2/ | 1·50
0·22 | <u>Ddt</u>

Ditto   (Piers

<u>Expansion joint</u>

5·47 | 25 'Flexcell' impregnated fibre board 150 deep designed joint to boundary including formwork to edge of reinforced concrete bed 225 deep

(F.7.2)

17

18

SUPERSTRUCTURE

Taking off list

Brickwork structure
Roof coverings, construction
    and rainwater goods
Steelwork and casing
Partitions
Internal Finishings
Windows

Brickwork Structure

Item		Bring to site and remove on completion all plant required for 'Brickwork and Blockwork' (G.2.1) & Maintain on site all plant required ditto (G.2.2)

Brickwork in load-bearing superstructure

(G.3.b)
External wall

Height from top of
slab to underside
parapet wall :

	3 000
	1 000
	525
	4525

Outer leaf
Trench girth	50 820
Less 2/300	600
	50 220

Corners 4/265	1 060
	51 280
Less 4/105	420
	50 860

Inner leaf
Trench girth	50 820
Less 2/300	600
	50 220
Corners Less 4/2/80	640
	49 580

GARAGE AND OFFICE

Load bearing
Superstructure (cont)

50·86	Half brick skin of
4·53	hollow wall in LBC
	'Tudor' facings in
	stretcher bond in
	gauged mortar
	(1:1:6) pointed
	with a neat flush
	joint as the work
	proceeds
	(G.14.9)
5·47	(Above
0·23	garage
	door

49·58	Ditto in LBC Flettons
4·53	in stretcher bond
	in gauged mortar
	(1:1:6)
	(G.5.3.c)
5·47	(Above
0·23	garage
	door

2/300   50 820
              600
          50220

50·22	Form 55 mm wide
4·53	cavity in hollow wall
	with galvanised
	twist pattern wall
	ties at the rate of
	5/m²
	(G.9.1)
5·47	(Above
0·23	garage
	door

(wall ties)

21

Piers

2/ 3·00

Attached pier in
LBC Flettons in
English bond
in gauged mortar
1500 wide × 215
deep projection

(G.5.4)

Crosswall
Height 4525
Less 225
        50   275
            4250

6·00	One brick wall
4·25	in LBC Flettons
	in English bond
	in gauged
	mortar (1:1:6)
	(G.5.3.a)

Note - fair face
to garage taken
with finishings

22

Brickwork in non-loadbearing Superstructure

<u>Parapet wall</u>

Girth	50 820
Less 2/300	600
	50 220
4/2/25	200
	5 470
	55 890

Height
4 Courses = <u>300</u>

55·89	
<u>0·30</u>	

One brick wall in LBC 'Tudor' facings in stretcher bond in gauged mortar (1:1:6) both sides pointed with a neat flush joint as the work proceeds including wall ties as before described at the rate of 5/m²

(G.14.9)

<u>55·89</u>	

Double splay header on edge coping in LBC 'Tudor' Standard special facings (BS 4729 Reference 2.4.6.1) bedded & jointed in gauged mortar (1:1:6) flush pointed on all exposed surfaces

(G.17.1)

4/ <u>1</u>

Extra for 90° angle (BS 4729 Reference 2.4.7.1)

20

23

24

Non-loadbearing
Superstructure (cont)

55·89	Extra over one brick wall in 'Tudor' facings for tile creasing set projecting comprising two courses of 254 wide asbestos cement slates laid breaking joint in gauged mortar (1:1:6) with mortar fillet to coping both sides (G.18.1)	

&

Close 55 wide cavity with asbestos cement slate. 200 wide laid horizontal in gauged mortar (1:1:6) (G.9.2)

Less ²/300	Gutter 50 820 600 50 220 5 470 55 690

55·69 0·40	'Hyload' pitch polymer horizontal cavity gutter lapped 150 at joints (no allowance made for laps) (G.37.3)

4/ 1	Extra for preformed 90° angle (G.37.3)

	Plinth course
Inner leaf	49 570 5 470 55 040

55·04	Extra over half brick skin of hollow wall in LBC Fletton brickwork for plinth course in Flettons (BS 4729 Reference 2.5.3.2) in gauged mortar (1:1:6) (G.17.1)

4/ 1	Angles (G.17.2)

Roof coverings, construction
and rainwater goods

Roof area
6 000
Less
²/250   500
5500
10000
8000
1900
1000
215
100
100
21 315

Item

Bring to site and
remove on
Completion all plant
required for
'Asphalt work'

(L.2.1)

&

Maintain on site
all plant required
for ditto

(L.2.2)

&

Bring to site and
remove on
Completion all plant
required for
'Roofing'

(M.2.1)

&

Maintain on site
all plant required
for ditto

(M.2.2)

21.32
5.50

50 Woodwool
showerproof quality
roof decking nailed
to joists at 400
centres

(M.30)

&

20 Asphalt to
comply with BS 988
in two coats in
flat coverings over
300 wide on and
including black
sheathing felt to
woodwool decking
(measured separately)

(L.4.3)

27

28

22

GARAGE AND OFFICE

Roof coverings etc (cont)

$$^2/_{100} \quad \begin{array}{r} 6000 \\ 200 \\ \hline 6200 \end{array}$$

$$400 \overline{\smash{\big)}\, 21\,135}$$
$$= 53 + 1$$
$$= \underline{54}$$

54/	6.20	50 × 225 'Protimized' sawn softwood flat roof joists exceeding 6.00 but not exceeding 6.30 long (N.2.1/N.1.6)
54/	5.50	50 wide 'Protimized' sawn softwood firrings average 75 deep (N.6.1)
2/	5.50	20 Asphalt skirting average 175 wide on face including two arrises, turn in of nib to groove (measured separately) and two coat angle fillet to roofing (L.6.1)
4/	1	Angles (L.6.1)

29

2/	21.32	20 Asphalt lining to gutter 500 girth on face including two arrises, turn in of nib to groove (measured separately) and two number two coat angle fillets (L.7.1)
2/	2	Ends (L.7.1)

$$\begin{array}{r} 21\,315 \\ 6\,000 \\ \hline 27\,315 \\ 2 \\ \hline 54\,630 \end{array}$$

	54.63	25 × 25 Horizontal chase in Fletton brickwork for turn-in of asphalt (G.11)
2/	21.32	50 Woodwool gutter boarding 200 wide laid to falls on and including 25 (average) thick 'Protimized' sawn softwood packing at 400 centres (N.4.1.f) & 50 Woodwool gutter side 100 wide (N.4.1.f)

30

23

		Rainwater Installation			
		(R.4.1.a)	2/	2	Cast iron holderbat for 76 diameter rainwater pipe cut & pinned to brickwork (R.15.2)
		Internally			
		(R.5.1.a)			
Item		Bring to site and remove on completion all plant required for 'Plumbing and Mechanical Engineering Installations' (R.2.1)			
		&	4/	2	Ditto including making good fair face (R.15.2)
		Maintain on site all plant required for ditto (R.2.2)	6/	1	Cast iron rainwater outlet for 76 diameter down pipe (R.14.1)
6/	4.50	76 Coated cast iron rainwater pipe to comply with BS 460 with open socketed joints fixed to brickwalls with holderbats (measured separately) (R.10.1)	6/	1	Cement and sand joint of 76 iron rainwater pipe to socket of 100 diameter clayware drain pipe (R.12.1)
		roof $\begin{array}{r} 3000 \\ 1000 \\ 150 \\ \underline{350} \\ \underline{4500} \end{array}$	6/	1	20 Asphalt lining to outlet for 76 diameter rainwater pipe (L.7.1)

GARAGE AND OFFICE

Builder's work in connection
with rainwater installation

(R.38.2)

6/	4·50	Prime and paint one undercoat and one finishing coat on pipework 150 - 300 girth internally (V.10.1)
4/	1	Hole in 25 'Asbestolux' ceiling for pipe not exceeding 0·30 girth (T.21.6)

Steelwork

Unfabricated steelwork

(P.3.1.a)

Item — Bring to site and
remove on
completion all
plant required
for Structural
Steelwork

(P.2.1)

&

Maintain on site
all plant required
for ditto

(P.2.2)

Item — Erect at a general
height of 4·00 above
ground level
steelwork having a
total weight of
_____ tonne

(P.10.1)

Preamble note — Steel to be
to BS 4 Part 1 Grade 43
(BS 4360)

33

34

25

Unfabricated steelwork (cont)
Entrance doors

	²/225	5470
		450
		5920

2/ 5.92

Beam at roof level
consisting of
254 × 102 × 28 kg/m
Universal beam 5.92
metres long
            (In No 2)

    × 28 kg/m = _____ kg
            (P. 4. 3)

    400 | 5920
         15 + 1
       = 16

16/ 2

Tapped hole through
6.4 mild steel for
12.7 diameter bolt
for Carpenter's fixing
            (P. 8.2.f)

            &

Ditto through 10
mild steel for 12.7
diameter bolt ditto
            (P. 8.2.f)

35

6.00

Beam at roof level
consisting of
305 × 165 × 54 kg/m
Universal beam
6.00 metres long
            (In No 1)

    × 54 kg/m = _____ kg

            (P.4)

8

M 12 bolt in mild
steel 75 long with
nut & washer and
connecting to
socket (measured
separately)
            (P. 8.4)

            &

Hole through 13.7
mild steel for 12.7
diameter bolt
            (P. 8.2.f)

36

26

2/	1

Precast concrete
(1:2:4 – 20 aggregate)
padstone 400 x 200 x
150 bedded and
jointed in gauged
mortar (1:1:6)
including four
cast-in sockets
(Harris and Edgar
HET 300 series) in
malleable iron 76
long to receive an
M 12 bolt (measured
separately)

(F. 18)

2/	6.00
	0.30
2/2/	6.00
	0.16
2/2/	5.92
	0.25
2/2/	2/5.92
	0.10

Paint one coat of
red oxide on
surfaces of steelwork
before fixing

(P. 9.1)

2/	6.00
	0.30
3/	6.00
	0.16

Touch up red oxide
and paint one
undercoat and
one full gloss
finishing coat of
paint on structural
metalwork surfaces
over 300 girth

(V. 6.1.b)

37

Casing to
steelwork

5.47

12.5 Exterior
quality plywood
fascia 390 wide
fixed to timber
framing
(measured separately)

(N.4.1.g)

&

Knot, prime and
paint two
undercoats &
full gloss finishing
coat paint on
general surfaces
of plywood
externally

Super x 0.400
= _____

(V. 4.1.g)

38

27

Casing to steelwork (cont)

2/ 5.47	50×50 Sawn Softwood bearers bolted (bolts and borings measured separately)
	(N.6.3)
16	50×50 Ditto 250 Long bolted (ditto)
	(N.6.3)
	(to soffite of Steel beams
2/2/ 16	Bore 50 thick Softwood for bolt
	(N.29.1.a)

39

Partitions
Non-loadbearing Superstructure

(G.3.1.c)

6.00 4.25 2.90 3.20 3.00 3.20	100 Precast concrete solid block partition to comply with BS 2028 Type 'B' size 440× 215 blocks bedded and jointed in gauged mortar (1:1:6)
	(G.27.1.a)
2/ 4.25 2/ 3.20	Bond end of 100 block partition to brickwall including forming pockets in brickwork and extra material for bonding
	(G.33)

Note - door adjustments made later

40

GARAGE AND OFFICE

Stud partitions

	3000
	100
	3100
3100	2900
100	750
3000	2150
door 750	
2250	

	2250	2150
Less 4/50	200	200
	2050	1950

75 x 50 Wrought
Softwood
partition

2/5/	3·10
	3·00
	2·90
	2·25
	2·15
4/	2·05
4/	1·95
2/2/	0·65

(N.2.1.b)

2/2/	3·10

Plug brickwork for
carpentry at 300
centres

(N.28)

	2·25
	2·15

Plug Concrete for
ditto at 300
centres

(N.28)

41

Internal Finishes

Item

Bring to site and
remove on
Completion all
plant required for
floor, wall & ceiling
finishes

(T.2.1)

&

Maintain on site
all plant for ditto

(T.2.2)

Ceilings

(a) Garage

10·00
6·00
5·47
0·27

25 Asbestolux
asbestos cement
board lining to
ceilings over 300
wide with butt joints
fixed to timber
joists at 400 centres
over 3·50 and not
exceeding 5·00 metres
high

(N.15.4)

42

29

Internal Finishes
Ceilings (cont)

(b) Office

Provide the P C Sum of
£ 550·00 for suspended
ceiling to be executed
complete by a Nominated
Sub- contractor

Add for profit

Allow for general attendance

Allow for special attendance
on nominated sub-
contractor executing
suspended ceilings
(approximately 66 m^2
total area) :-

Moveable scaffolding,
unloading,
distribution, hoisting
& placing in position

Power supply maximum
load 1·5 kW

(B.9.3)

51·80	
3·00	

2/ 3·00
3·00

0·10
3·00

Wall finishes

(a) Office

8000	
6000	
14000	
×2	
28000	28000
3000	
2900	
5900	
×2	
11800	11800
6000	
×2	
12000	12000
	51 800

15 Carlite plaster
in two coats to
brick or block walls
over 300 wide

(T.5.1/T.4.2)

(Partition
wall

Ditto not
exceeding 300
wide

(T.4.2)

(End of
partition
wall

43

44

30

Wall finishes (cont)

(a) Office (cont)

2/	3.00	Expamet angle bead to wall plaster on brick or block walls (T.5.7)
	28.00 3.08	Extra over Fletton brickwork for grooved bricks (G.8)
4/	0.10 3.08	Ddt Ditto (Partition intersection)
	3.00 3.00 2.90 3.00	9.5 Gyproc plaster board wall lining fixed to timber with galvanised clout headed nails and taping joints (N.15.4) & 3 Skim coat of setting plaster on plasterboard walls (T.4.2)

45

	3.00 3.00 2.90 3.00	3 Skim coat on plasterboard walls as before but in compartments not exceeding 4.00 m² (T.3.2)

2/1000
2.90	3.00
2.00	2.00
4.90	5.00

	5.00 3.00 4.90 3.00	Ddt 15 Carlite to brick or block walls as before & Add Ditto but in compartments not exceeding 4.00 m²
2/ 2/ 2/	3.00 3.00 2.90 3.00 51.80 3.00 3.00 3.00 0.10 3.00	Prepare and apply one mist coat & two full coats emulsion paint on plasterboard walls over 300 wide - internally (V.4.1.a)

46

31

Wall finishes (cont)
(b) Garage

	10 000
	6000
	10 000
2/265	530
2/2/215	860
	27 390

27.39	Extra over Fletton
4.00	brickwork for fair
	face and flush
	pointing as the work
	proceeds

(G.14.3)

2/2/ 0.22	Ddt
1.00	Ditto

(Pier less
than
wall
height

47

Floor finishes
(a) Garage

10.00	Trowel surface of
6.00	set concrete bed
5.47	
0.27	(F.9.3)

&

Apply one coat of
Johnson's 'Monotax'
to surfaces of
trowelled concrete
floor

(F.9.3)

(b) Office

8.00	50 Cement and
6.00	sand (1:3)
	trowelled bed
	laid on concrete
	base over 300 wide

(T.13.2)

3.00	Ddt
0.10	Ditto

(Partition

48

GARAGE AND OFFICE

Floor finishes (cont)

(b) Office (cont)

8·00 6·00	Marleytex CRX plain carpeting laid on trowelled bed (T. 29.2)
	Ddt
3·00 0·10	Ditto

(c) Toilets

2·90 1·90 3·00 1·90 2/ 0·75 0·10	50 cement and sand (1:3) trowelled bed laid on concrete base over 300 wide (T. 13.2) & 2·5 x 300 x 300 'Marley flex International' vinyl asbestos tile flooring laid on trowelled bed over 300 wide (T. 14.1)
2·90 1·00 3·00 1·00	<u>Add</u> last <u>two</u> items but in compartments not exceeding 4·00 m² on plan (T. 3.2)

49

Windows

10	Softwood casement to comply with BS 644 to suit opening 2100 x 1200 overall in three lights, two opening, each light in two panes; comprising 55 open rebated casements, 75 x 135 twice rebated, grooved and twice moulded jambs; 75 x 135 twice rebated, moulded, chamfered & throated head; 100 x 135 twice rebated and four times moulded mullion and 100 x 225 twice splay rebated, moulded, grooved and throated sill as Drawing JK/1 (see Provisional Sum for ironmongery) (N. 21)

Provide the Provisional Sum of £100 for ironmongery

(A. 8. 1. a)

50

Windows (cont)

10/2/	2	3.2 × 75 Galvanised mild steel fixing cramp 225 girth once bent with one end screwed to softwood and other end built into brickwork

(N.31)

$$\begin{array}{r} 2100 \\ 1200 \\ \hline 3300 \\ \times 2 \\ \hline 6600 \end{array}$$

10/	6.60	Bed wood frame in gauged mortar (1:1:6) and point one side in non setting mastic

(G.43.3)

$$2/75 \begin{array}{r} 2100 \\ 150 \\ \hline 2250 \end{array}$$

10/	2.25	25 × 135 Wrought softwood window board with one edge rounded and other edge tongued to groove in sill (groove measured separately)

(N.13.1.f)

10/	2	Notched returned ends (N.14.3)

51

8/6/	0.56 0.50	6 OG Quality clear sheet glass and glazing to wood sashes with sprigs and putty in panes 0.10 – 0.60 m²

(U.4.1)

Toilet window

2/6/	0.56 0.50	4 Obscure glass Stippolyte pattern and glazing ditto 0.10 – 0.50 m²

(U.4.1)

10/	2.10 1.20	Knot, prime, stop and paint one undercoat and one full gloss finishing coat paint on surfaces of timber windows divided into medium panes – internally

(V.5.1.d)

&

Ditto – externally

(V.5.1.d)

10/2/2/	1.10	
10/2/2/	0.60	Opening edge – externally

(V.5.1.e)

52

34

Windows (cont)

Adjustment for openings

Inner leaf 1800 x 1200
Sill 75
Lintel 225 _____ 300
                 1800 x 1500

Outer leaf 1800 x 1200
Sill 75
Lintel 75 _____ 150
                 1800 x 1350

10/	1.80	1.50

Ddt
Half brick skin of hollow wall in Flettons in gauged mortar (1:1:6)

10/	1.80	1.35

Ddt
Half brick skin of hollow wall in LBC 'Tudor' facings in gauged mortar (1:1:6)

&

Ddt
Form 55 wide cavity

53

4/	1.80	1.50

Ddt
Extra over Flettons for fair face
(Garage

6/	1.80	1.50

Ddt
Extra over Flettons for grooved bricks
(Office

&

Ddt
15 Carlite plaster to brick or block wall

&

Ddt
Emulsion paint to plastered walls

2/	1.80	1.30

Ddt
15 Carlite plaster to block wall in compartment not exceeding 4.00 m²

&

Ddt
Emulsion paint to plastered walls

10/2/	1.20

Fair return on 'Tudor' facings not exceeding half brick wide

(G.14.10)

54

35

10/2/	1·20	**Adjustment for openings** (cont')

Close 55 cavity at
jamb of opening
with Fletton brickwork
half brick wide
including inserting
Hyload pitch polymer
damp proof course
103 wide (no
allowance for laps)

(G.9.2/
G.37)

(Garage

| 4/2/ | 0·16
1·20 | Extra over Fletton
brickwork for fair
face as before |

(G.14.3)

(Office etc

| 6/2/ | 0·16
1·20 | Extra over Fletton
brickwork for
grooved bricks |

(G.8)

| 4/2/ | 0·15
1·20 | 15 Carlite plaster to
walls not exceeding
300 wide |

(T.4.2)

| 2/2/ | 0·15
1·20 | Ditto but in
compartments not
exceeding 4·00 m² |

(T.3.2)

| 6/2/ | 1·20 | Expamet plaster
bead |
| 6/ | 1·80 | (Lintel |

(T.5.7)

| 6/ | 1·80 | Expamet plaster
stop bead |

(T.5.7)

55

Heads & Sub-sills

2/150   1800
      300
      2100

| 6/ | 1 | 225 x 215 x 2100 |

Precast reinforced
concrete (21 N/mm² –
20 aggregate) boot
lintel with fair face,
soffite and front
edge of toe and
reinforced with
three 10 mild
steel bars, lintel
bedded and
jointed in gauged
mortar (1:1:6)

(F.18)

| 4/ | 1 | 225 x 215 x 2100 |

Ditto with fair
face back, soffite
and front edge
of toe ditto

(Garage

(F.18)

| 10/ | 2·10
0·36 | 'Hyload' pitch
polymer gutter
(no allowance for
laps) |

(G.37.3)

| 10/ | 2 | Extra for end
cloak piece |

(G.37.3)

56

GARAGE AND OFFICE

Windows (cont)

Heads and subsills (cont)

6/ 1   300 x 69 x 2100
Precast reinforced
concrete (21 N/mm² -
20 aggregate)
weathered and twice
grooved sill with
fair face soffite and
front edge; reinforced
with two 10 mild
steel bars, sill
bedded and jointed
in gauged mortar
(1:1:6)   (F.18)

4/ 1   300 x 69 x 2100
Ditto with fair face
to back, soffite and
front edge, ditto

    (F.18)

10/ 2.10   4 x 20 Galvanised
mild steel water
bar bedded in
cement and sand
(1:1)
    (Q.7.2)

4/ 1.80
   0.15   15 Carlite plaster
to soffite of concrete
lintel not
exceeding 300 wide
    (T.4.2)

57

2/ 1.80
   0.15   15 Carlite plaster
to soffite of
concrete lintel
not exceeding
300 wide in
compartments
not exceeding
4.00 m²

    (T.5.3/T.3.2)

Protection

Worker up - include
protection items
at the end of
each work
section

    (F.45 etc)

End of Garage & Office Example

58

37

Single storey structure 2

Project: Common room, Stanmore Street

Drawings: Common room and boiler house
plan and part section   (HDA 154 03/7A)

Common room sections   (HDA 154 03/8)

Common room elevations   (HDA 154 02/12)

General arrangement community
centre foundations and roof   (HDA 154 SE1/5)

Common room roof truss   (HDA 154 SE1/6)

Generally

This is the second example of a single storey structure which extends the range of items covered in the first example. The common room is in the centre of a housing estate to be built at the same time.

Column 3

The reference to trial hole information and water table level can be assumed to have been described or defined elsewhere in other documents relating to the project.

Column 4

The area of the hexagon is taken as $L_1 \times a + L_2 \times b$ where:

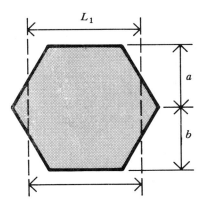

L_1 and L_2 are adjusted for the differing projections due to changes in foundation detail. The existing ground is assumed to be level. The excavation for reduced levelling is taken to the underside of the hardcore.

The edges of the excavation are sloped in order to keep the soil well away from the external wall and thus in danger of bridging the d.p.c. It is not an item of any great significance and could well be omitted.

Column 5

As the trenches are in the main backfilled with hardcore, the soil from the trenches is measured as removed from site, and the hardcore filling measured as backfilling. Adjustments will be made later for earth backfilling.

Column 6

The lengths of the trench excavation for the lower three sides of the pentagon need to be adjusted for the pits at the corners (1200 × 1200 × 1800 deep). These pits are shown on plan but not clearly identified on the sections. Although the sides of the trenches are shown as being splayed, they are measured as being vertical and for the width of the concrete foundations therein. If sloping sides were really necessary (they are assumed in this case only to be a draftsman's style) they would be measured in accordance with D.42.1. The working space is needed for the edge of the slab. Hardcore filling is not measured as the hardcore base blinding and concrete slab are measured later (see Columns 19, 21, 22).

Column 7

When measuring the earthwork support it is a good idea to shade in on the plan the sides as the dimensions are booked. In this way adjustments or timesings which need to be made can be seen. At the corners the adjustment is as follows.

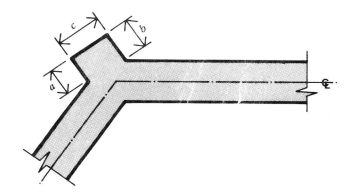

Earthwork support to sides *a* and *b* are covered by the over-measurement which occurs at the centre line (the earthwork support is measured to twice the centre line length). It is only necessary to add in for face *c*. The dimension for additional length of support to the projection to the sides of the centre trench is taken as 0.30 m. The passings occur at the left and right hand sides of the centre trench where the perimeter abuts.

Column 8

The rule for measuring pits can be difficult to interpret when the pit is a deepening or widening of a trench. A rough guide is that when the pit is in a deeper depth classification then it is measured as a pit and not as part of the trench. The taker-off should use his discretion when measuring and if it is a significant volume then the approach adopted should be described in the bill of quantities.

Column 9

Measurement of excavations to soft spots and removing obstructions is a good example of a taker-off identifying items which may occur on site but which do not appear on the drawings. Quantities for these items (which are marked as Provisional to highlight the need for adjustment once the extent of the work is known) should be assessed realistically by reference to the total volume of excavation and the nature of the site.

Column 11

The approximate quantity of concrete will be inserted by the worker-up. Specification details are taken from elsewhere.

Column 13

The projection of the one brick wall to both ends of the centre wall are measured as attached piers in Column 15.

Column 14

The preamble note is included to eliminate repetition in the measured items. The cavity is measured the full height.

Column 16

The facings are measured on the centre line of the exposed face, hence ½/0·255 mm or 127·5 mm adjustment for each corner. The fraction 4/6 is the proportion at each corner. It would be 6/6 if it were a right angle. The height of 225 is based on 150 (the floor slab thickness) and 75 below existing ground level.

Column 18

The trenches are measured as being excavated and then backfilled with hardcore. The volume of hardcore being displaced by the walls is now deducted.

Column 20

The slab to the boiler room is a suspended slab and formwork (left in) will need to be taken to the soffite. There is no need for blinding but 200 mm hardcore filling is measured in the void between foundation level and the underside of the slab.

Column 21

Note the requirement (not necessary in this example) of clause F.4.6 to describe beds laid on earth or hardcore.

Columns 25 and 26

The concrete beds are reinforced with fabric mesh whilst the boiler room slab is reinforced with bars. The quantities entered in Column 26 can be assumed to have been taken from an engineers provisional bending schedule. (For a detailed comment on bending schedules see Example 4). The reinforcement would be remeasured when carried out.

Column 28

It is debatable whether the concrete to detail S—S should be measured as concrete in foundation trenches or as a bed. In this example it has been measured as in foundation trenches. An addition 75 mm is added to the bed for the thicknessing. Rule F.5.3 states that the defined thickness as rule F.5.2 ignores any such projections. Thus although at this point the bed becomes 150 mm + 75 mm = 225 mm thick it is still defined as not exceeding 150 mm thick.

Column 30

Provisional reinforcement is taken as in Column 26.

Column 31

The first calculation determines the projection 'p' of the foundation beyond the face of the wall as shown below:

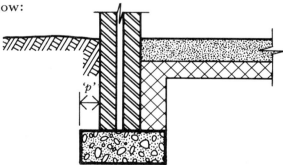

Column 32

Although items such as Protection are often included in the Bill at a later stage by the worker-up it is good practice for the taker-off to include the items in his measurement.

Column 33

The external walls are measured first followed by the internal walls. Within the heading of external walls the top half of the hexagon is taken first and then the two piers to the bottom half are measured.

Column 37

The brick wall is measured in the following sequence:

1. Central length to top side of joist to truss (deducts of brickwork for concrete beam on section B—B to take with beam).
2. Two triangular shapes above concrete beam.

Facings are taken to the net area (i.e. not including the area covered by the skirting). The deduction for the fireplace is taken in Column 39. The 1000 mm width of the chimney breast is taken from the dimensions on drawing SE1/5.

Column 38

There is a difficulty in interpreting clause G.4.1 in that brickwork should be measured by the average height. If one looks at the wall, the top is sloping. Does this imply the following?

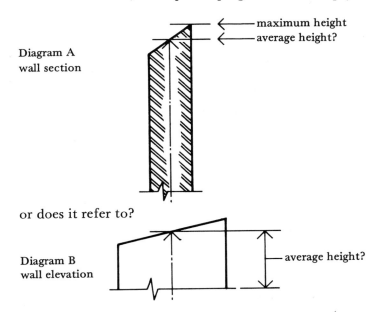

Diagram A
wall section

or does it refer to?

Diagram B
wall elevation

In which case the height in the above section is taken as the maximum height.

43

The method adopted in this book (and unless the item has a significant quantity it is more of an academic than real issue) is to measure the average height as defined in Diagram A and measure rough cutting to form a chamfered angle as clause G.10. Care must be taken in calculating the area of brickwork and the following approach is taken.

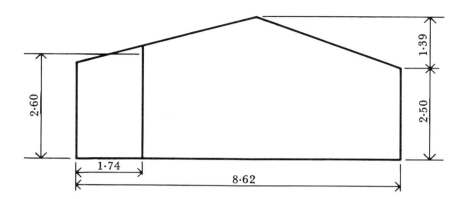

An adjustment is made for the lobby where the wall is built in facing bricks. The height of 2.60 is taken as the mean height.

Column 42

It is thought most convenient to measure the beams with the structure but they could alternatively be picked up when the window adjustments are made.

Column 44

The variation in profile to the ends of the three beams (Section V—V) are enumerated, notwithstanding clause F.15.5. A qualification to this effect is included in the taking off for subsequent inclusion in the bill. It is felt that the linear measurement stated in clause F.15.5 is inappropriate to this adjustment.

Column 45

The area on each face of the roofing tiles is calculated as follows:

base width   : 6·50 (scaled from drawing SE1/5)
sloping height : 6·07 (based on formula in dimensions; 5·60 is the scaled length on plan of the centre of the triangular face from drawing SE1/5. The height of 2·34 is taken from drawing 03/8.

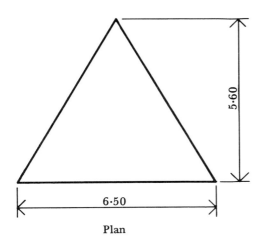

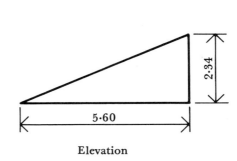

Plan	Elevation

Column 46

The length of the hips is calculated as:

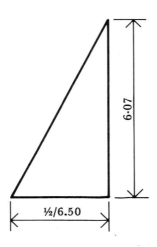

Column 47

The lead saddle is measured as a full value item as it can be fixed as well as supplied by the plumber. Only in the case of soakers and slates which have to be made by the plumber and fixed by the tiler should two items (i.e. supply and fix) be measured.

Column 48

The item for the capping is not very satisfactory but it is likely to be adequate for pricing purposes unless at construction stage a detailed drawing is provided which indicates more work than can be envisaged at this stage. In the event of a complex cap being needed the item would be omitted and a new item measured and priced.

45

Column 49

The rafter lengths need to be calculated from the plan of the roof (on drawing SE1/5). First, the pitch of the roof is calculated. This can be done either by using an adjustable set square and finding the angle from the section on drawing SE1/6 or by trigonometry. Using the drawing SE1/6 (section on top left hand corner) the trigonometric calculation is:

$$\text{Tan } \theta = \frac{1934}{1520 + 1520 + 2362}$$

$$= \frac{1934}{5402}$$

$$= 0.3580$$

therefore

$$\theta = 20^{\circ}$$

Now for each rafter length, x, the formula is derived from

$$\text{Cos } 20^{\circ} = \frac{\text{Plan length}}{x}$$

therefore

$$x = \frac{\text{Plan length}}{\text{Cos } 20^{\circ}}$$

Therefore for the central rafter, which has a plan length of 5.65

$$x = \frac{5.65}{0.94}$$

$$= 6.00$$

The other rafter lengths are similarly calculated. With the cosine being constant in each length, the calculation is easy. A side note of the methodology used is adequate.

Column 50

The length of the wall plate is the outside length, the bedding is the centre line length.

Column 51

The purlins which are left exposed and then stained need to be described as such. The exposed faces of the purlin are wrought. The half brick wall is taken from the architect's detail.

Column 54

Although not shown on the drawing, connectors are taken between rafters and purlins and also at the ends of the purlins where they fix onto the hip rafters.

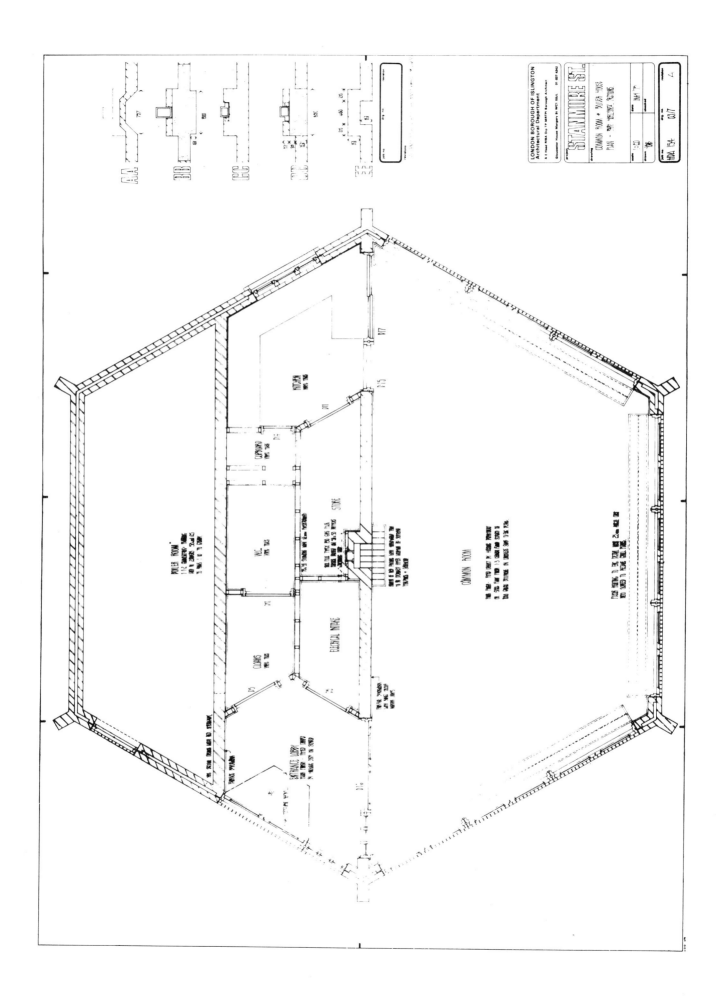

LONDON BOROUGH OF ISLINGTON
Architectural Department

Gloucester House Margery St WC1 9SJ

STANMORE ST

project
COMMON ROOM & BOILER HOUSE
PLAN - 1:50 HEATING SECTIONS

scale 1:50   date MAY '87

drawn by   checked

job no. drg no.   revision
HRA 154   03/7   A

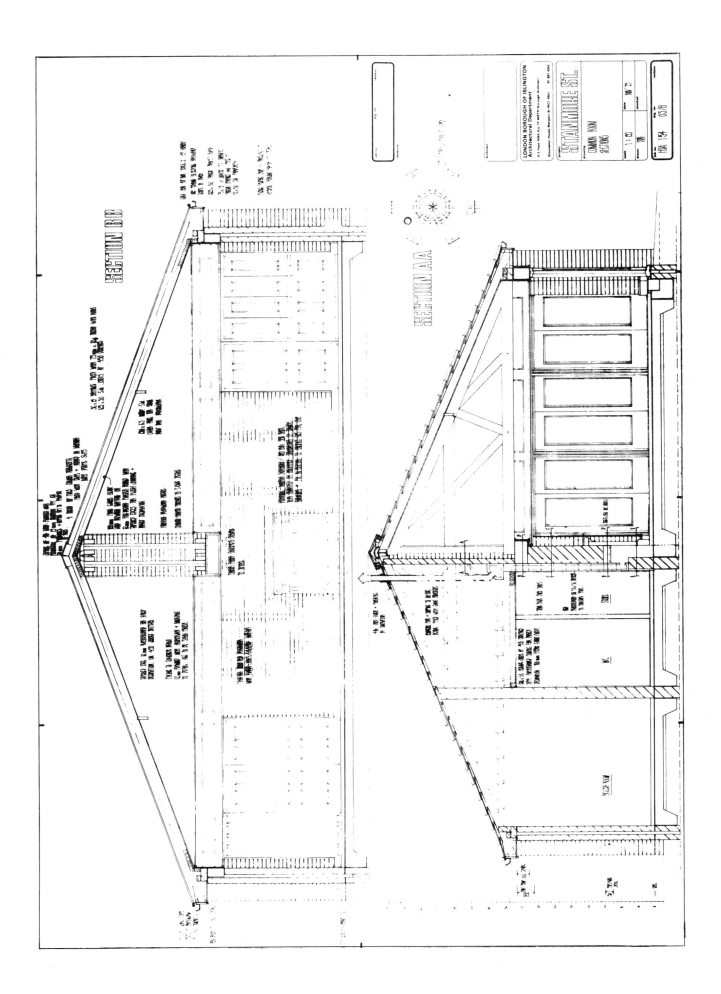

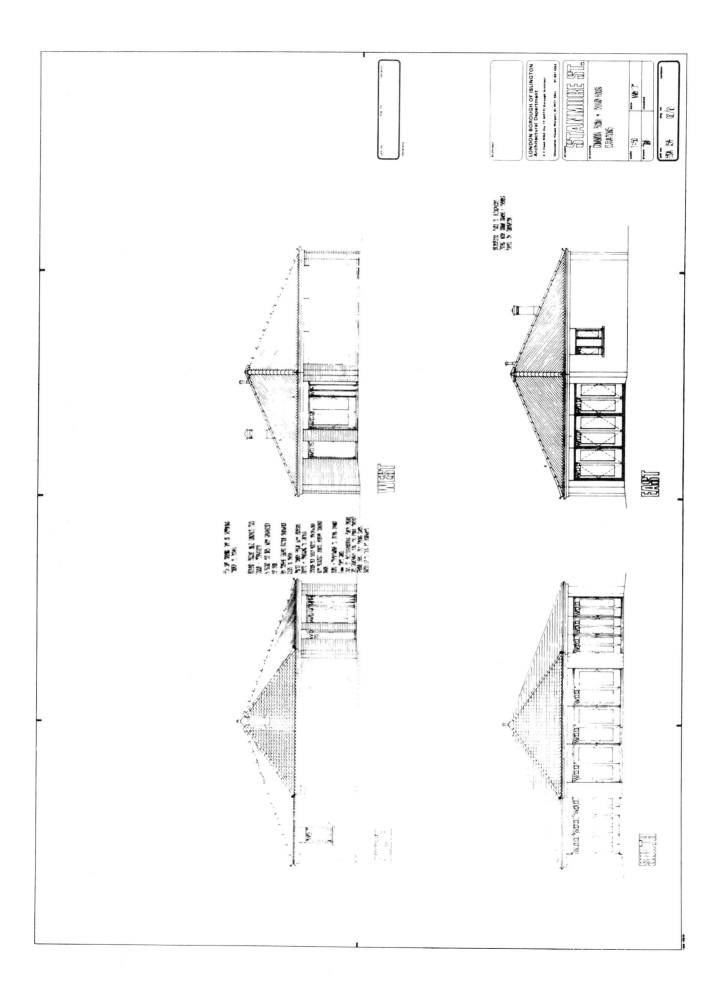

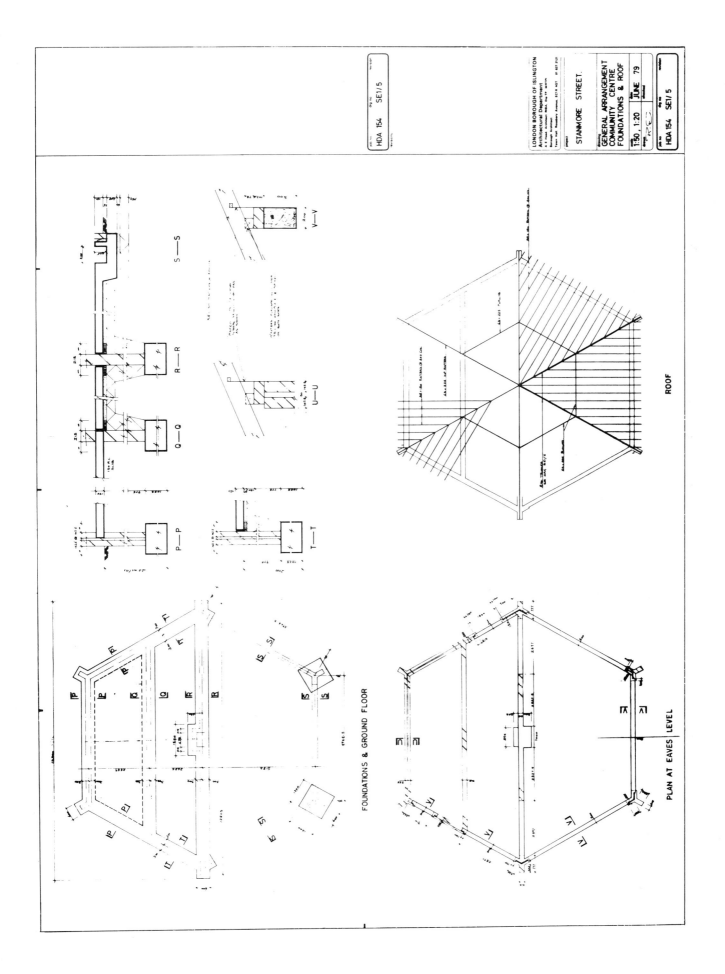

LONDON BOROUGH OF ISLINGTON
Architectural Department
A E Head Architect RIBA, Dip TP MRTPI
Borough Architect
Town Hall Rosebery Avenue, EC1R 4QT   01 837 2121

STANMORE STREET.

GENERAL ARRANGEMENT
COMMUNITY CENTRE
FOUNDATIONS & ROOF

1:50, 1:20   JUNE   79

HDA 154   SE1/5

HDA 154   SE1/5

S — S

V — V

U — U

R — R

Q — Q

P — P

T — T

ROOF

FOUNDATIONS & GROUND FLOOR

PLAN AT EAVES LEVEL

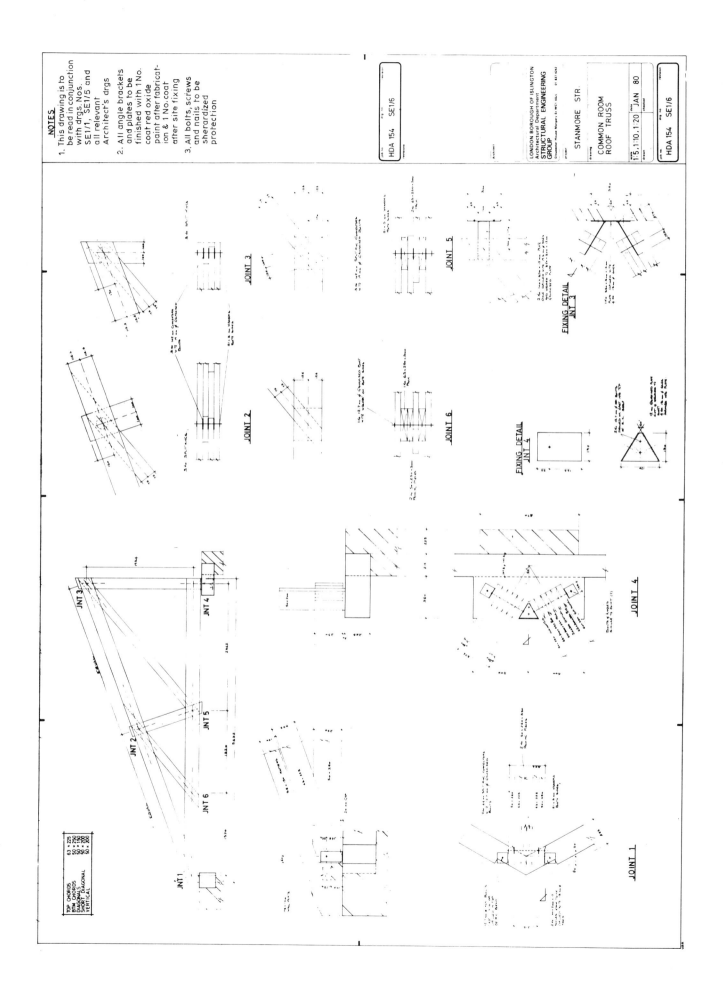

HDA 154   SE1/6

LONDON BOROUGH OF ISLINGTON
Architectural Department
STRUCTURAL ENGINEERING
GROUP

STANMORE STR.

COMMON ROOM
ROOF TRUSS

1:5,1:10,1:20   JAN   80

HDA 154   SE1/6

JOINT 3

JOINT 5

JOINT 2

JOINT 6

FIXING DETAIL
JNT 3

FIXING DETAIL
JNT 4

JOINT 3

JOINT 4

JOINT 1

JNT 3
JNT 4
JNT 5
JNT 2
JNT 6

JNT 1

TOP CHORDS	63 × 225
BTM CHORDS	50 × 250
DIAGONALS	50 × 200
SHORT DIAGONAL	50 × 200
VERTICAL	

STANMORE STREET

COMMON ROOM AND BOILER ROOM A

Taking-off List

SUBSTRUCTURE

Nature of ground
Plant
Oversite excavation
Trench excavation
Pit excavation
Water & hard dig
Concrete in foundations
Brickwork
Hardcore
Oversite concrete
Damp proof course
Edge detail
Earth backfill
Protection

BRICKWORK SUPERSTRUCTURE

External walls
Piers
Internal walls
Chimney breast/stack
Ring beam
Protection

ROOF

Coverings
Linear items
Rafters, purlins etc
Trusses
Metalwork
Eaves construction
Rainwater installation
Protection

2

53

SUBSTRUCTURE

The foundation layout and details are shown on Drawing SE1/5

(D.2)

The Contractor is referred to the trial hole information given in the Preambles to 'Excavation and Earthwork' for the nature of the ground

(D.3.1.b)

Ground water was found at 23.00 above sea level on 10 April 1981

(D.3.1.a)

Item	Bring to site and subsequently remove from site all plant required for the 'Excavation and Earthwork' section
	(D.4.1)

Item	Maintain all plant for this section of the work
	(D.4.2)

3

Oversite excavation

```
              13 200          5965
  2/332        664   2/300     600
              12536           6565
                             12536
                          2 |19101
                             9551

              5183            2624
  ½/200        100            2553
              5283             300
                             5483
```

9·55	Excavate to reduce levels maximum depth not exceeding 0·25 m
5·48	
9·15	× 0·20 = _____
5·28	(D.13.3)
4/ 0·60	&
0·33	Remove excavated material from site
2/ 0·60	× 0·20 = _____
1·20	(D.29)

```
150
 50
200
```

6/ 5·97	Finish sides of excavation to reduce levels to approximately 45° slope
0·25	
4/2/ 0·33	
0·25	
2/2/ 0·90	
0·25	(D.42.1)

4

COMMON ROOM

Trenches
(Top half)

centre	13200
Sides	5965
"	5965
"	5365
middle	8400
	39495

Depth
2000
Less 200
1800

(Bottom half)

5365
Scaled 600
" 400   1000
    4965

5365
Scaled 2/600   1200
    4765

Depth
275
200
50
150
475
Less 2·00
275

39.50
0.60
1.80

4/ 0.60
   0.60
   1.80

2/½/ 0.60
     0.30
     1.80

1.50
0.35
1.80

Excavate foundation trench over 0.30 m in width maximum depth not exceeding 2.00 m starting at reduced level

(D.13.6.b)

&

Remove excavated material from Site

(D.29)

&

Hardcore filling to excavations

(D.35)

650
Less 300
350

2/ 4.97
   1.00
   0.28

4.77
1.00
0.28

Excavate foundation trench over 0.30 m in width maximum depth not exceeding 0.25 m starting at reduced level

(D.13.6.b)

&

Remove excavated material from Site

(D.29)

2/ 4.97
   0.25
   0.28

4.77
0.25
0.28

Excavate working space maximum depth not exceeding 0.25m starting at reduced level and fill with hardcore

(D.12.1.a)

5

6

Trenches (cont)

2/2/	4·97
	0·28
2/	4·77
	0·28

Earthwork support maximum depth not exceeding 1.00 m, distance between opposite faces not exceeding 2.00 m

(D.17)

2/	39·50
	1·80
4/	0·60
	1·80
2/	0·30
	1·80
2/	0·35
	1·80

Ditto maximum depth not exceeding 2.00 m, distance between opposite faces not exceeding 2.00 m

(D.17)

4/	0·60
	1·80

Ddt
   Ditto

(Passings

7

	39·50
	0·60
4/	0·60
	0·60
2/½/	0·60
	0·30
	1·50
	0·35
2/	1·20
	1·20

Level and compact bottom of excavation to receive hardcore or concrete

(D.41)

(Concrete footing

(Pits

Pits

2/	1·20
	1·20
	1·80

Excavate pit for isolated foundation base to pier with maximum depth not exceeding 2.00 m & having both dimensions less than 1.25 m on plan

(In No 2)
(D.13.5)

&

Remove excavated material as before
(D.29)

&

Hardcore filling as before (D.35)

2/	
4/	1·20
	1·80

Earthwork support maximum depth not exceeding 2.00 m, distance between opposite faces not exceeding 2.00 m as before

(D.17)

8

56

COMMON ROOM

General items

Item	Keep the surface of the site and the excavations free of Surface water	
	(D.25)	

Item	Protect the work in the ' Excavation and Earthwork' section	
	(D.45)	

Soft spots

5.00 1.00 1.00	Excavate to remove soft spots maximum depth not exceeding 4.00 m commencing at reduced level (Provisional) (D.13)

&

Remove excavated material from site
(Provisional)
(D.29)

&

Plain insitu lean mix (1:12 – 40 aggregate) in foundations to trenches over 300 thick
(Provisional)
(F.6.2)

obstructions

5.00 1.00 1.00	Extra over trench excavation for breaking up existing brickwork (Provisional) (D.13.12)

&

Extra over trench excavation for breaking up existing concrete
(Provisional)
(D.13.12)

9

10

Concrete

Other concrete work

Note The approximate volume of insitu concrete in this substructure is
———— m³

(F.3.2)

39·50	Sulphate resisting cement concrete (1:6 – 40 aggregate) foundations in trenches over 300 thick poured against face of excavation
0·60	
1·03	
2/ 0·60	
0·40	
1·03	
2½/ 0·60	
0·30	
1·03	
1·50	
0·35	
1·03	

(F.4.6/ F.6.2)

&

Ddt
Hardcore filling to excavation

(D.35)

2/ 1·20	Concrete as before described in isolated foundation bases over 300 thick poured against face of excavation (In No 2)
1·20	
1·03	

(F.6.3)

&

Ddt
Hardcore filling to excavation

(D.35)

Item   Bring to site and subsequently remove from site all plant required for the 'Concrete Work' section

(F.2.1)

Item   Maintain all plant for this section of the work

(F.2.2)

11

12

58

COMMON ROOM

Brickwork

Item	Bring to site and subsequently remove from site all plant required for the 'Brickwork and Blockwork' section (G.2.1)
Item	Maintain all plant for this section of the work (G.2.2)

Brickwork in foundations
(G.3.1.a)

External wall

Outer skin   3/5365
  19035   = 17895
(See Column 16)   2/400 = 800
Inner skin
  18695   Centre line   18695
  340
  18 355   Height   975
         150
         1125

18·36	
1·13	Half brick skin of hollow wall in blue engineering bricks in sulphate resisting cement mortar (1:3) (G.5.3.c)
19·04	
0·90	
2/2·60	
0·15	
2/2/ 0·32	
0·98	
2/2/ 0·32	
0·15	

Less   975
        75
(Pits) 900

(0·15 in Section T-T Inner skin)

13

All half brick thick walls are in stretcher bond. All one brick walls and over are in English bond.

18·70	
1·13	Form 50 (nominal) cavity between brick skins of hollow wall including bonding the skins together with galvanised mild steel vertical twist type wall ties to comply with BS 1243 Table 3 spaced 900 apart horizontally and 450 apart vertically including keeping cavity free from mortar droppings and rubbish (G.9.1)
0·32	
1·13	

975
150
1125

2/ 1·13	Close 50 (nominal) cavity of hollow wall at ends of wall with blue engineering brickwork half brick thick (G.9.2)
2/2/ 1·13	

14

		Foundation brickwork (cont)

18·70
0·98
0·05
2/2/ 0·32
0·98
0·05

Sulphate resisting cement concrete (1:10 - 20 maximum aggregate) filling to cavity of hollow wall not exceeding 100 mm wide

(F.16.8/ F.5.2)

6/ 0·98

Projection of attached pier in blue engineering bricks in sulphate resisting cement mortar (1:3) 215 wide and 332 mm projection

(G.5.4)

```
              12803
   332
   255
2/587       11 74
            11 629
```

15

11·63
1·13

9·00
1·13

2/ 0·35
1·13

One brick wall in blue engineering bricks in sulphate resisting cement mortar (1:3)

(Recess in centre wall
(G.5.3.a)

Preamble Note

Facing bricks are Ibstock Multi-red stock facings and pointing with coloured mortar with raked joints

Outer leaf in facings

```
                18695
4c/4/½/255        340
                19035
```

19·04
0·23

Half brick skin of hollow wall in facings in coloured sulphate resisting cement gauged mortar(1:1:6) built fair and pointed one side

(G.14.9)

```
   864
    85
   949
   150
    75
   225
```

19·04
0·23

(bottom half of hexagon

16

COMMON ROOM

Foundation
brickwork (cont)

18.70		Sulphate resisting concrete filling to cavity of hollow wall as before
0.23		
0.05		

(bottom part of hexagon

6/	0.23	Projection of attached pier in facing bricks in coloured sulphate resisting cement gauged mortar (1:1:6) 215 wide and 332 projection built fair and pointed one face and both returns

(G. 14. 9)

17

6/2/	0.23	Fair birdsmouth angle formed by cutting facings

(G. 15. 3)

Adjustment of Backfill

Trench depth	1800
Less Concrete	1025
	775

18.70		Ddt
0.26		
0.78		Hardcore filling to excavations
6/	0.33	
	0.23	
	0.78	
21.33		
0.23		
0.78		

	11.63
	9.00
2/350	0.70
	21.33

18

61

Hardcore Bed

12803

²/₃₃₂ 664

12139

5965
255
255
6475
12139
2 | 18614
mean = 9307

9.31
5.18
___

(bottom

150 Bed of hardcore
laid level and
blinded to receive
concrete including
levelling and
compacting
surface of excavation

(D.36/D.40)

2624   12139
225   9000
2399   2 | 21139
            10570

(top
middle
5964.5
²/₃₅₀ 700.0
5264.5

(fireplace

10.57
2.40
___
0.68
0.35
___

3/5.26
0.25
___

(Extra
for Detail
S-S -
slope

1.11
0.35
___

Ddt

Ditto   (Chimney
              breast

19

7.30
2.32
___

3/ 5.26
___

Hardcore filling
200 thick to
make up levels
under suspended
slab construction

(D.36)

2559
112
127   239
       2320
       5800
       8800
    2 | 14600
       7300

150
50
200

Handpack
hardcore to
form battering
face 250 high
and blind to
receive concrete

(D.38)

(Detail
S-S

20

COMMON ROOM

Concrete Beds

9·31 5·18 0·15	Reinforced sulphate resisting cement concrete (21 N/mm² – 20 maximum aggregate) bed not exceeding 150 thick laid level (F. 6. 8)
10·57 2·40 0·15	
0·68 0·35 0·15	
	Note – edge detail S-S adjusted later

	Ddt
1·11 0·35 0·15	Ditto (Chimney breast

Blinding Bed

(adjustment for 200 wide flexcell at edges)

```
      5183
       200
      4983

      2400
2/200 = 400
      2000

      10570
  2 =  400
      10170
```

9·31 4·98 0·05	Sulphate resisting cement concrete (1:10 – 20 maximum aggregate) in 50 thick binding bed laid level (F. 6. 8)
10·17 2·00 0·05	
0·68 0·35 0·05	

	Ddt
1·11 0·35 0·05	Ditto (Chimney breast

```
2310    7300
 317     205
2627    7505
```

7·51 2·63 0·15	Reinforced sulphate resisting cement concrete (21 N/mm² – 20 maximum aggregate) suspended bed not exceeding 150 thick laid level (F. 6. 9)

7·30 2·31	Formwork to soffite of horizontal reinforced concrete slab left in (In No 1 surface) (F. 15. 1)

Concrete Beds (cont)

2/	2·60	Half brick skin of
	0·15	hollow wall in
		blue engineering
		bricks as before
		(Detail T-T
	8·70	Ddt
	0·15	One brick wall in
		blue engineering
		bricks as before
		(Detail Q-Q
	12·14	50 Expanded
2/	10·57	polystyrene board
		SD grade
2/	2·40	compression joint
		strip 200 wide
2/	0·35	laid level on
		hardcore
		(F.7.2)

&

10 'Flexcell' expansion joint 150 wide placed vertically between edge of concrete and brick wall including temporary support while bed is cast

(F.7.2)

23

Dampproof course in concrete

	9·31	'Bituthene 1000'
	5·18	self adhesive
	10·57	preformed
	2·40	membrane from
		W R Grace Ltd
	0·68	laid level on
	0·35	concrete and
		lapped a minimum
	7·51	of 100 at all joints
	2·63	(no allowance
		made for laps)
		(G.37.2)
	1·11	Ddt
	0·35	Ditto
		(Chimney breast

24

64

COMMON ROOM

Concrete Beds (Cont)

Reinforcement

2/	9·31 5·18	Steel fabric reinforcement Reference A 142 (200×200 mesh) weighing 2·22 kg/m² in floating beds lapped 150 all edges		
2/	10·57 2·40			
2/	0·68 0·35			

(F.11.4. b/
F. 12.2)

2/	1·11 0·35	Ddt Ditto	(Chimney breast)

2/2/	5·60	Raking cutting on ditto
2/2/	2·70	

(F.12.6)

See Structural
Engineers
Provisional Quantities
for steel in
suspended slab

25

High yield steel

0·05 Tonne	16 Straight and bent bars in suspended slab (Provisional) (F.11.4.b)
0·20 Tonne	12 Ditto (Provisional)
0·30 Tonne	8 Ditto (Provisional)

26

Brickwork
Dampproof course

2/	18.70	
2/	2.60	

'Hyload' pitch polymer horizontal damp proof course 102 wide lapped 150 at joints on brickwork (no allowance made for laps)

(G. 37. 2)

	11.63
	9.00

Ditto but 215 wide

(G. 37. 2)

(recess in centre wall

2/	0.35
6/	0.33

27

Edge detail S-S

```
              5965
Less ⅔/332    664
              5301
              800
              900
          2 | 1700
              850
```

3/	5.30
	0.85
	0.20

Reinforced sulphate resisting cement concrete (21 N/mm² - 20 maximum aggregate) foundations in trenches not exceeding 300 thick poured against face of excavation

(F. 6. 2)

3/	5.30
	0.32
	0.08

Reinforced concrete bed not exceeding 150 thick laid level as before

(F. 6. 8)

(Thicknessing to bed at recess

3/	5.30
	0.10
	0.23

Reinforced concrete as before described in walls in sides of horizontal duct in bed not exceeding 100 thick

(F. 6. 12. d)

28

COMMON ROOM

Edge detail S-S (cont)

3/	5·30	Formwork to edge of foundation bed not exceeding 250 high (F.14.1) (Outer edge
3/	5·30 0·25	Ditto to sloping top face of blinding bed over 15° from horizontal (In No 3 areas) (F.15.3)
3/	5·30	Formwork to give a fair face finish to face of concrete foundation and edge of bed not exceeding 250 high (F.14.1) (to edge of duct
3/2/	5·30 0·23	Ditto to vertical face of wall (In No 6 surfaces) (F.16.1)
3/2/	0·23	Ditto to end of wall 75 wide

29

See Structural Engineer's Provisional Quantities

High yield steel

	0·12 Tonne	16 Straight and bent bars in ground slab (Provisional) (F.11.4.b)
	0·10 Tonne	12 Ditto (Provisional)
	0·05 Tonne	10 Ditto (Provisional)
	0·10 Tonne	8 Ditto in walls (Provisional)

Adjustment on bed

2/3/	5·30 0·30	Ddt Steel fabric reinforcement Reference A142 as before
3/	5·30 0·30 0·15	Ddt Reinforced concrete bed not exceeding 150 thick as before

30

Adjustment
for earth
backfilling
to outside
face of wall
_____

width

foundation width   600
Less   wall   255
        2| 345
           172 5

length
      5965
      5965
      5965

Sides of projection
   8/150   1200
ends ditto   4/600   2400
short returns
to lower half
   2/250   500
        21995

31

22.00
0.18
0.78
_____

2/2/   0.50
0.70
0.78
_____

2/   0.22
0.18
0.78
_____

Item

Item

Ddt
Hardcore filling to excavation as before

     &   (Piers

Add
Backfill approved selected excavated material around foundations
       (D. 35)

     &

Ddt
Remove surplus excavated material as before

Protection

Protect the work in 'Concrete Work' section
      (F.45)

Protect the work in 'Brickwork and Blockwork' section
      (G.58)

Note to Worker - up
These two items apply to Superstructure as well as Substructure Section Bill

32

COMMON ROOM

SUPERSTRUCTURE

Load bearing Superstructure

           (G. 3. b)

      External walls

    Centreline   18 695
  ½/ 4½/256   340
              19 035

19.04	Half brick skin of	
2.50	hollow wall in	

facings in coloured
mortar (1:1:6)
built fair and
pointed one side

        (G. 14. 9)

18.70	Form 50 (nominal)
2.50	cavity between

brick skins of
hollow wall all as
described in
'Substructure'

        (G. 9. 1)

33

18.36	Half brick skin
2.50	of hollow wall

in Flettons in
gauged mortar
(1:1:6)
     (G.5.3. c)
       18 695
         340
       18 355

    Boiler Room

11.80	Extra over Fletton
2.50	brickwork for

fair face and
flush pointing
     (G. 14.3)
       5.800
  2/3 000 6 000
       11 800

2/2/	2.50

Fair cut
birdsmouth
angle on Fletton
brickwork

     (G.15.3)

34

External walls
(Cont)

2/	2·50	Projection of attached pier in Flettons in gauged mortar (1:1:6) 215 wide and 150 projection (G.5.4) (Centre wall
4/	2·50	Projection of attached pier in facings in gauged mortar (1:1:6) 215 wide and 332 projection built fair and pointed one face and both returns (G.5.4)
2/	2·50	Close 50 cavity of hollow wall at jambs of opening with Fletton brickwork half brick thick (Against screens (G.9.2)
2/2/	2·50	(Piers added back

35

2/	2·50	Fair return to facing brickwork to reveals of openings half brick wide (Piers (G.14.10)

Piers

2/	1·03 2·50	Half brick skin of hollow wall in facings in gauged mortar (1:1:6) as before (G.14.9)
2/	0·86 2·50	Form 50 wide cavity as before (G.9.1)
2/	0·69 2·50	Half brick skin of hollow wall in Flettons in gauged mortar (1:1:6) as before (G.5.3.e)
2/	2·50	Projection of attached pier in facings 215 wide and 332 projection as before
2/ 4/	2·50	Fair cut birdsmouth angle on facings (G.15.3)

36

70

COMMON ROOM

Internal Walls

Central Wall

(Measured net – no openings to take)

	2/2528	5056
		1000
		6056

2½/	6·06	One brick wall in Flettons in gauged mortar (1:1:6)
	2·70	
	5·40	(G.5.3.a)
	1·95	

	6·06	Extra over Fletton brickwork for facing bricks built fair and pointed one side
	2·00	
	1·00	(G.14.3)
	2·50	

	1·00	One brick wall in Flettons in gauged mortar (1:1:6) in chimney breast projection
	2·25	(G.5.4)

To take   Deducts of Wall for concrete beam

37

Boiler Wall

	8·62	One brick wall in Flettons in gauged mortar (1:1:6)
	2·50	
2½/	8·62	(G.5.3.a)
	1·39	(Above Lobby Wall

	8·50	Extra over Fletton brickwork for fair face and flush pointing
	2·50	
½/	8·50	(G.14.3)
	1·39	(In Boiler Room

Lobby

Ddt	1·74	One brick wall in Flettons as before
	2·60	

&

Ddt   Extra over for fair face as before

&

Add   One brick wall in facings in gauged mortar (1:1:6) and flush pointed both sides

(G.14.9)

38

71

Support Stack,
Hearth and Flue
and Fireplace
adjustment

0.90	Half brick wall in
2.05	facings in gauged
	mortar (1 : 1 : 6)
	built fair and flush
	pointed one side
	(G.14.9)
	(A-A

2/ 2.40	Fair birdsmouth
	angle
	(G.15.3)
	(A-A

0.80	Half brick wall in
0.20	Flettons in gauged
	mortar (1 : 1 : 6)
	built against
	concrete including
	grouting solid at back
	(G.5.3.a)

0.45	Ddt
0.90	One brick wall as
	before
	(B-B

1.50	Ddt
1.20	Extra over for
	facings as before

39

Provide the Provisional Sum
of £ 100.00 for collecting,
renovating and fixing
in position old cast iron
fireplace and marble
surround

(A.8.1.a)

Precast units

1	150 x 327 Reinforced
	concrete (21 N/mm² -
	20 maximum
	aggregate) lintel
	890 long with and
	including three
	12 mild steel bars
	and finished fair
	on soffite and
	bedded and
	jointed in gauged
	mortar (1 : 1 : 6)
	(F.18.1)
	(over fireplace
	&

1800 x 215 x 225
thick overall
reinforced concrete
padstone with
800 long x 100
wide x 225 thick
projection on two
opposite faces with
and including
two 12 mild
steel bars and
bedded and
jointed ditto
(F.18.1)
(Section B-B

40

72

COMMON ROOM

<u>Stack (cont)</u>

Precast dense refractory
concrete 'Type 150' gas
flue blocks bedded and
jointed in 'flue joint'
refractory mortar as
manufactured by.......

(G.54)

2.60	Standard units (Reference 5×1)
1	[Extra for single flue block with front entry (Reference 5×3)
	&
	Extra for angle conversion unit (Reference 10×2)
	(See Mechanical Installation for flue pipe and capping

41

<u>Ring Beam</u>

(Detail V – V)

```
      2597
²/100  200
       910
       200
      1360
       200
      ─────
      5467
```

(Central Wall)

```
        2597
²/102·5  205
        ─────
        2802
```

5·47 0·20 0·30	Reinforced concrete (21 N/mm² – 20 maximum aggregate) isolated beam over 0·03 but not exceeding 0·10 m² sectional area
²/2·80 0·20 0·30	(In No 5) (F.16.3)

```
        2597
²/102·5  205
        ─────
        2802
```

5·48 0·30	Ddt Half brick wall in skin of hollow wall in commons
²/2·80 0·30	Ddt One brick wall in commons as before

42

73

		Ring beam (cont)		5.47	Formwork to sides & soffite of plain rectangular isolated horizontal beam on wall 300 deep and 200 wide
		See Structural Engineer's Provisional Quantities	2/	2.80	
0.27 Tonne		16 High yield steel straight and bent bars in isolated beams (Provisional) (F.11.4.j)			(In No 5 members) (F.17.1/F.15.5)
			3/2/	1	Extra over ditto for variation in profile to 150 wide face for a length of 100
0.07 Tonne		12 Mild steel links and stirrups in isolated beams (Provisional)			
0.25 Tonne		8 Ditto (Provisional)			Note for Bill Notwithstanding the provision of Clause F.15.5 Variations in profile to the ends of beams have been enumerated
					3/5965 = 17895
				17.90	Horizontal brick-on-edge eaves course one brick wide in Flettons in gauged mortar (1:1:6) (G.21.1)
				2	Irregular internal angle (G.21.2)
				Item	Protect all the work in this section (G.58)
					To take – plywood facing to concrete beam
		43			44

COMMON ROOM

Roof

Coverings

Item	Bring to site and subsequently remove from site all plant required for the 'Roofing' Section (M.2.1)	

Item	Maintain all plant for this section of the work (M.2.2)	

Note to Worker-Up

Repeat above items for Roof Construction

6/½/ 6.50
    6.07 | 418 x 332 Interlocking Redland Regent Mark 2 concrete roofing tiles laid to 20° pitch to even courses with 75 laps fixed with Clips every fourth course to and including 19 x 38 impregnated sawn softwood battens nailed to timber rafters with 10g galvanised nails (M.5/M.6)

$$x^2 = 5.60^2 + 2.34^2$$
$$= 36.83$$
$$x = 6.07$$

45

6/½/ 6.50
    6.07 | Inodorous reinforced bituminous felt to comply with BS 747 Type IF underfelting lapped 150 at joints nailed to timber rafters at 450 centres (M.17.1)

$$x^2 = \frac{6.5^2}{2} + 6.07^2$$
$$= 47.40$$
$$x = 6.88$$

6/2/ 6.88 | Raking cutting underlay (M.17.2) & Cutting roofing tiles to hips subsequently covered (M.11.1)

6/ 6.50 | Purpose made Clips to eaves tiling

6/ 6.88 | Extra over tiles laid to 20° pitch for purpose made third round hip tiles fixed with Clips and bedded in coloured cement mortar (1:3) including dentil slips in each pan (M.11.3)

6/ 1 | Ends filled solid with coloured cement mortar (M.11.3)

46

Coverings (cont)

6/1

25 x 6 Galvanised
steel hip iron 450
long once bent
and twice countersunk
drilled and screwed
to softwood
(M.13)

Capping

1

Hole for small pipe
through roof tiling
(M.7.4)

&

Code No 4 lead
cover saddle to
hexagonal timber
cap to sloping
roof 850 x 850
(extreme) over
apex between
opposite corners,
dressed to profile
of cap, over hip
tiles and turned
into tile roofing
including soldered
joints as required
and hole for
small pipe

(M.55)

47

1

19 Thick marine
quality plywood
Grade 3:3
hexagonal shaped
cap to sloping
hexagonal roof
700 x 700 (extreme)
over apex between
opposite corners
including raking
and splay cutting
and traversed
for lead and on
and including
50 x 125 sawn
softwood packing
pieces and firring
pieces

To take — hole in
roof for boiler
flue to take
with boiler
installation

48

Rafters, Purlins etc

Rafters

6/2/ 3.90	
6/2/ 3.20	38 x 150 Impregnated sawn softwood carcassing member in pitched roof
6/2/ 2.50	(N.2.1.d)
6/2/ 1.80	
6/2/ 1.10	
6/2/ 0.40	

6/2/ 4.60 38 x 150 Ditto in 4.50 – 4.80 lengths (N.1.6)

6/2/ 5.30 38 x 150 Ditto in 5.10 – 5.40 lengths (N.1.6)

6/1/ 6.00 38 x 150 Ditto in 5.70 – 6.00 lengths (N.1.6)
(Centre rafter

Note – Rafter lengths calculated as plan lengths ÷ Cos 20° (0.94)

49

Plate
5965
75
6040
6
36240

36.24 75 x 100 Impregnated sawn softwood carcassing member in kerbs and bearers (N.2.1.e)

6/ 5.96 Bed plate 100 wide in gauged mortar (1:1:6) (G.43.1)

Hips

6/ 6.90 63 x 225 Impregnated sawn softwood carcassing member in pitched roof (N.2.1.d)

50

77

Rafters, purlins etc (cont)

Purlins

5/ 3·00	63 × 225 Sawn softwood carcassing member impregnated with a clear preservative and stained light brown before fixing in pitched roofs
	(N.2.1.d)

2/225   450
        63
       513

4/ 3·00	Wrought surface 513 girth
	(N.3.2)

Eaves Detail
(Detail U–U + V–V)

6/ 5·96	Horizontal brick-on-edge course one brick wide in Flettons in gauged mortar (1:1:6) set back 40 mm on top of hollow wall
	(G.21.1)

6	Irregular internal angle
	(G.21.2)

6/ 5·96 0·25	Half brick wall in Flettons in gauged mortar (1:1:6)
	(G.5.3.a)

51

Trusses
Composite Items

2	Wrought softwood impregnated with a clear preservative and stained light brown before fixing in monopitch roof truss 5632 × 2280 overall comprising 63 × 225 top chords, 50 × 250 bottom chords, 50 × 150 ties, 50 × 200 struts and 50 × 200 verticals all as shown on the accompanying drawing SE1/6 with necessary timber connectors, bolts and angle cleats (Metal connection plates fixing to the structure at joints 1, 3 and 4 and rawlbolts and drilling the structure have been measured separately)
	(N.18)

Preamble Note - All the following metalwork to be finished one coat red oxide paint after fabrication and a second coat after fixing on site

1	'Fixing detail JNT 3' as shown on Drawing SE1/6 comprising 12 thick mild steel face plate 350 × 300 four times drilled for 12 diameter bolt welded to two 12 × 100 × 370 mild steel fixing plates each having one end drilled for 19 bolt
	(N.31.1)

52

COMMON ROOM
Trusses (cont)

1	'Fixing detail JNT 4' as shown on Drawing SE1/6 comprising 12 mild steel truncated triangle shaped base plate 170 × 130 overall twice drilled for 12 bolts and welded to two 150 × 200 face plates each drilled for 12 bolt
	(N. 31. 1)
2/2/1	12 Mild Steel angle bracket 100 wide × 300 girth once bent and twice drilled for 12 bolt at joints 1 and 4
	(N. 31. 1)
8	Drill reinforced concrete beam or padstone for 12.7 rawlbolt fixing and grout with cement mortar (1:1)
	(F. 9. 9)
4	Drill brickwork for ditto and grout with cement mortar (1:1)
	(G. 50)

Preamble Note   All bolts, screws and nails to be sheradized

2/ 2	12 × 89 Loose bolt type Rawlbolt (Reference G.20)
4	
4	
	(N. 31. 1)

Carpenter's Metalwork
(Drawing SE1/6 Details U-U, V-V)

6/ 15	4.6 × 25 'Mafco' galvanised steel strap 685 girth once bent, one end twice counter-sunk drilled and screwed to softwood and other end split forked and built into joint of brickwork
6	
	(N. 31. 1)
	(feet of rafters and hip end
5/ 7	'Mafco' dual clip rafter to purlin connector
5/ 2	
	(N.31.1)

Eaves Construction

Length of eaves

$$\begin{array}{r} 5965 \\ 2/250 \quad \underline{500} \\ 6465 \\ \underline{\times 6} \\ 38790 \end{array}$$

38·79	32 × 225 Wrought softwood fascia
	(N.4.1.g)

$$\begin{array}{r} \& \quad 225 \\ \underline{380} \\ 605 \end{array}$$

Knot, prime, stop and paint two undercoats and one finishing coat gloss paint on general surfaces of woodwork 150-300 girth - externally

(V.3.2)

6	Mitres on 32× 225 fascia
	(N.1.10)

55

38·79	19 × 75 Wrought softwood boarded eaves soffite kept clean for subsequent staining 400 wide with vee tongued and grooved joints, one edge tongued to and including groove in fascia (bearers measured separately)
	(N.4.1.g)

6	Notch 19 boarded eaves soffite around brick pier 889 girth
	(N.1.10)

38·79	Knot and apply two coats light brown stain to general surfaces of woodwork - externally
0·38	
	(V.4.1.g)

56

80

COMMON ROOM

Eaves Construction
(cont)

6/	17	50×50 Impregnated sawn softwood first fixings in eaves bearers not exceeding 300 long
		(N.6.3)
6/17/	0·40	50×50 Impregnated sawn softwood first fixings in eaves bearers
6/	5·80	(N.6.3)
6/	5·80	Plugging brickwork for carpentry fixing
		(N.28)

To take – fibreglass quilt and plaster-board ceiling finishes to main roof
— timber ceiling, plasterboard and insulation over cloaks, WC, store etc as Drawing 03/8

Rainwater installation

38·79	116×51 'Key Terrain' black UPVC standard square gutter (Reference 2250) jointed at 4·0 m centres with joint bracket (Reference 2251) twice screwed to softwood fascia and fixed at 1·0 m centres with support brackets (Reference 2252) once screwed to softwood fascia
	(R.7.1)
6	Extra over for 120° angle (Reference 2254)
	(R.7.3)
2	Extra over for running outlet (Reference 2253) twice screwed to softwood fascia
	(R.7.3)

Rainwater Installation (cont)

2/ 2·65	58 × 58 'Key Terrain' black uPVC standard square rainwater pipe (Reference 2200) jointed at 2·50 m centres with connector (Reference 2207) and fixed at 2·00 m centres with pipe and fitting clip (Reference 2213) twice plugged and screwed to brick wall (R.10.1)	
2	Connect 58 × 58 pipe to 110 × 110 vitrified clay gully with cement and sand (1:1) trowelled smooth to slope including wire baffle (R.12.1)	

Item	Mark the position of holes, mortices, chases and the like in the structure as required (R.37.2)
Item	Test the whole of the foregoing rainwater installation in accordance with the local sanitary authority requirements, remedy any defects and leave the whole in sound and proper working order to the satisfaction of the Architect and Local Authority (R.37.5)
Item	Bring to site and subsequently remove from site all plant required for the 'Plumbing' section (R.2.1)
Item	Maintain all plant for this section of the work (R.2.2)

59

60

COMMON ROOM

Protection Of Roof Elements

Roofing

Item	Protect the work in this section
	(M.56)

Woodwork

Item	Protect the work in this section
	(N.33.2)

NB – there is no timber flooring

Plumbing

Item	Protect the work in this section
	(R.41)

End of Common Room Example

61

Section Two
Specialist Structural Elements

Underpinning

Project: 6 Monk Street, Reading
Drawing: Underpinning to basement wall   (PG/82 UP/1)

Generally

Underpinning is not an everyday occurrence but it occurs frequently enough to make it worthwhile including in this book. This example has also been adapted and somewhat contrived to illustrate in particular the problem of excavating below the water table level. The project is an existing high street terraced shop and office premise which is to be modernized. The basement floor to ceiling height needs to be increased to provide space for printing machinery which will be used by the company renting the office space on the first and second floors. The party and front walls do not need underpinning as they extend below the new basement floor but the central wall does require underpinning.

Column 2

The requirements of clause H.1 are straightforward. It is necessary for the contractor to have details of the structure to be underpinned (i.e. not just the underpinning details) so that he can price the requirements of clause H.3.1 (temporary supports to structure).

Column 5

The width of trench is determined by the maximum depth of excavation. In this case the working space (i.e. the total width of trench) is 1·50 m. No deduction is made for the trial pit nor for the projecting foundations and footings.

A calculation needs to be made to see whether any of the earthwork support should be classified as next to roadway (clause D.19).

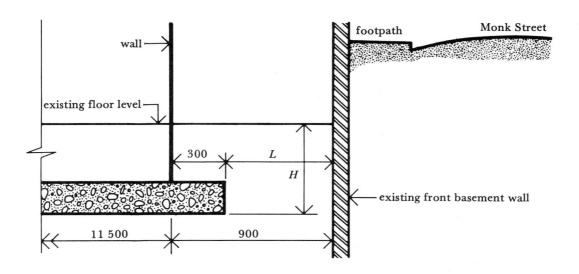

$L = 900-300$
$L = 600$
$H = 950$

As $L<H$, describe the support as next to roadway (see practice manual). For the purpose of calculating L the footpath is taken to the inside of the basement wall.

Column 7

A difficulty exists over the measurement of working space. As stated by Goodacre* it is difficult to see how clause H.3.2 can be compatible with H.3.3. The example below will illustrate this point

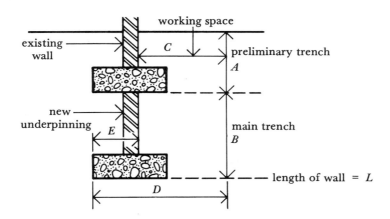

* Goodacre, P.E. (1979). Standard Method of Measurement: An introduction to the 6th edn. College of Estate Management, Reading.

The preliminary trench is calculated as $L \times C \times A$ (which includes working space). The main trench is calculated as $L \times D \times B$ (which also includes working space). Therefore the measurement of separate working space under clause D.12 cannot exist. The only way it can exist is if the working space is taken as $L \times C \times B$ and a trench is taken for $L \times E \times B$. This is an extremely unrealistic approach and is not recommended. It is possible that working space could occur at the ends of the wall to be underpinned as shown below.

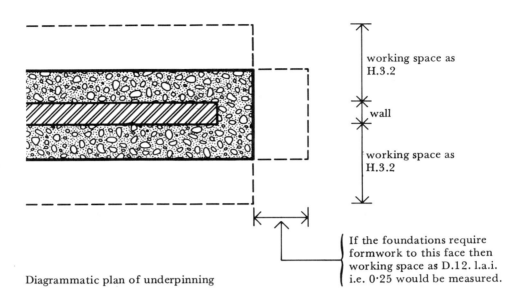

Diagrammatic plan of underpinning

working space as H.3.2

wall

working space as H.3.2

If the foundations require formwork to this face then working space as D.12. l.a.i. i.e. 0·25 would be measured.

Again this seems a very theoretical approach and it is therefore suggested that a clause be inserted along the following lines in the Bill:

'Notwithstanding clause H.3.3 of the Standard Method of Measurement, all working space has been included in the volumes of excavation of preliminary and main trenches as defined by clause H.3.2. Items of backfilling, disposal of soil, earthwork support, etc. have been measured to these volumes or areas. However the provisions of clause H.3.2 in respect of not adjusting working space shall still apply.'

Column 10

Formwork is taken to both long sides of the foundation but is not needed at the ends as it can satisfactorily be cast against the earth face. This also applies to the ends of the short lengths in which the concrete is cast.

Column 12

Clause D.13.12 requires that breaking up be given extra over the various types of excavation. In this case it is considered reasonable to group the preliminary and main trenches together.

Column 13

The item for surface water is defined in the Practice Manual as being intended solely for rain-water. As this is in an existing building the problem is not likely to arise but it is better to measure it and assume the contractor will not price it.

Column 14

The 'End of the following' is required so that the worker up knows that any further items are not to be billed under that heading. In this case it is also the end of the example and is obviously not essential as it would be if other measured items (such as the formation of the new basement floor) followed on.

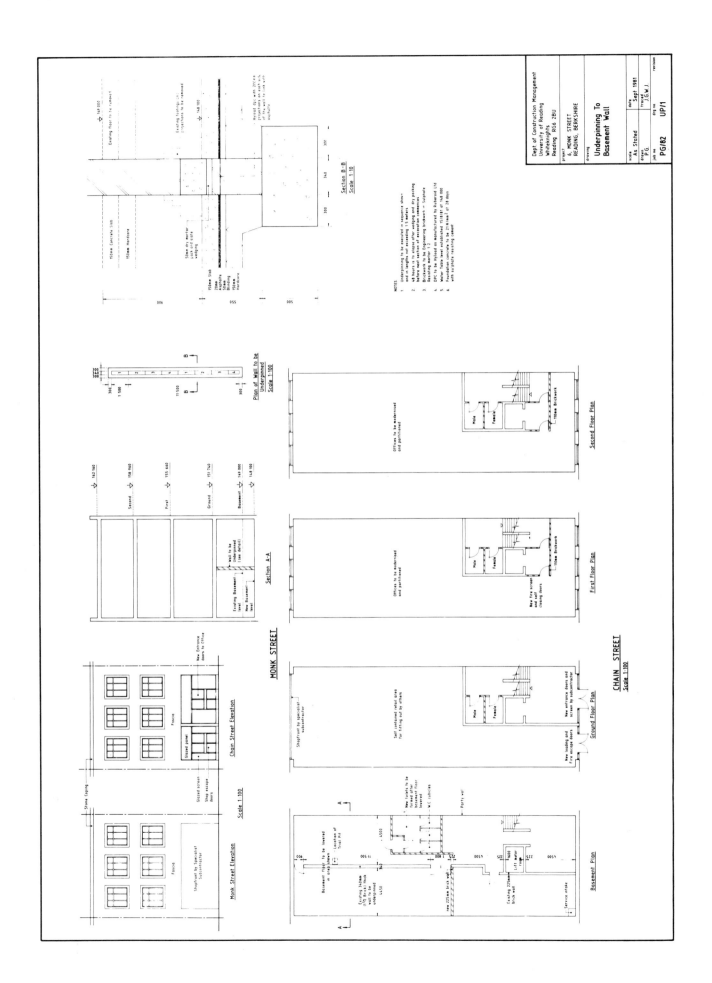

MONK STREET - READING

UNDERPINNING

Taking-off list

1 Description of the work

2 Trial pit and water table level

3 Plant

4 Temporary support to wall

5 Preliminary trench

6 Foundation trench (part below water table level)

7 Concrete

8 Brickwork

9 Water items

10 Protection

The following in underpinning basement wall 11.5 m long in lengths not exceeding 1.5 m with not more than two sections exposed at any one time as shown on Drawing UP/1.
The work is all within an existing building and may be carried out from both sides

(H.1.1/₂,₄,₅)

$(H.1.1/_{2,4,5})$

A trial pit 2 m deep (to underside of proposed new foundation) has been dug adjoining the wall to be underpinned

(H.1.3)

1

2

The ground water level
was established on
15/8/81 at 148.000

(H.1.3)

Item | Bring to site and
Subsequently remove
all plant required
for this section of
the work

(H.2.1)

Item | Maintain on site
all plant required
for this section of
the work

(H.2.2)

3

Item | Provide temporary
Support to wall
to be underpinned
11.5 m long and
4.04 m high
above top of
existing foundation

(H.3.1)

151 740
148 100
3 640

Less ground
floor slab   300
                 3 340

Plus below
basement
        Slab   700
                 4 040

4

MONK STREET - READING

The following in underpinning (cont)

Preliminary trench

2/	12.10	Excavate preliminary trench to base of existing foundations not exceeding 1.00 m deep Starting at existing basement level
	1.50	
	0.95	
2/	0.34	(H.3.4.a)
	0.30	
	0.95	

&

Remove surplus spoil from site

(D.35)

	length		depth	
	11 500		149 000	
2/300	600		148 100	
	12 100		900	
			050	
			950	

maximum depth
900
550
500
1950

5

2/	12.10	Extra over preliminary trench excavation for breaking up 150 thick concrete bed and 150 thick hardcore under
	1.50	
2/	0.34	(D.13.11)
	0.30	

2/1 500	3000
	340
	3340

2/	12.10	Earth work support to preliminary trench not exceeding 1.00 m deep and not exceeding 2.00 m between opposing faces
	0.95	
	3.34	
	0.95	

(H.5.1.a)

	3.34	Ditto and next to roadway
	0.95	

6

Main Trench

2/ 12.10
1.50
1.00

2/ 12.10
0.34
1.00

Excavate foundation trench starting below base of existing foundations not exceeding 1.00 m maximum depth

(H.3.4)

&

Backfill excavation with selected excavated material

(D.35)

depth
500
550
1050
Less   50
1000

2/ 12.10
1.50
0.95

3/ 12.10
0.34
0.95

Extra over for excavating below the ground water level

(D.13.13)

148 000
1050
underside of foundations 147 050

148 000
147 050
950

7

12.10
0.94

Level and compact bottom of excavation to receive concrete

(D.41)

2/300   600
340
940

2/ 12.10

Cut away projecting foundation consisting of two courses of brick footings and 200 wide concrete foundation 250 thick

(H.3.5)

2/ 12.10
1.00

Earthwork support below the level of existing foundations maximum depth not exceeding 1.00 m and exceeding 2.00 m and not exceeding 4.00 m between opposing faces and extending below water table level

(H.5.1.b)

8

94

MONK STREET - READING

The following in underpinning (cont)

Main Trench (cont)

Ends & sections

7/ .2/	3.34	Earthwork support
	1.00	as before described

but not exceeding 2.00 m between opposing faces

2/1 500   3000
                   340
                3 340

3.34	Ddt
1.00	Ditto

&

Add

Ditto and next to roadway

9

Concrete

12.10	Plain concrete
0.94	(21 N/mm² with
0.50	Sulphate resisting

cement) foundation in trench over 300 thick

(F.6.2)

2/ 12.10	Formwork to

edge of plain concrete foundation over 250 thick and not exceeding 500 thick

(F.14.1)

Brickwork

50   550
50   100
        450

11.50	One and a half
0.45	brick thick.

(340) wall in engineering bricks in sulphate resisting cement mortar (1:2) in English bond

(G.5.3.a)

10

95

11.50	Prepare the underside of existing 340 wide concrete foundation to receive the pinning up of new work
	(H.3.6)
	&
	Wedge and pin up new brickwork to underside of existing foundation 340 wide with Slates and 50 thick mortar dry pack
	(G.44)

$$\frac{2}{200} \frac{\begin{array}{r} 340 \\ 400 \end{array}}{740}$$

11.50 0.74	Hyload horizontal damp proof course bedded in sulphate resisting cement mortar and lapped 150 at side and end laps (measured net)
	(G.37)

11

12.10 0.94 1.00	Ddt Earth backfilling
	&
	Add Remove surplus spoil from site
	(D.29)

1.00 1.00 1.00	Extra over all excavation in underpinning for breaking up brickwork (Provisional)
	(D.13.12)
	&
	Ditto but concrete (Provisional)
	(D.13.12)

12

MONK STREET - READING

The following in underpinning
(cont)

Water

End of the following
in underpinning

Item | Keep the surface
of the site and
the excavations
free of surface
water

(D.25)

Item | Keep the excavations
below the water
table level free
of ground water

(D.26)

Protection

Item | Protect the work
in this section

(H.7)

End of underpinning example

13

14

97

EXAMPLE FOUR

Reinforced concrete

Project: Brookside Place Block A
Drawings: 1st Floor R.C. Details   (HDA 114 SE/A3 A)
                R.C. Details of staircase
                Ground — 1st   (HDA 114 SE/A4 A)
                Bending schedules   (HDA 114 0A3/01)
                                        (HDA 114 0A3/02)
                                         (HDA 114 0A3/03)
                                         (HDA 114 0A4/01A)

Generally

This is a small area of reinforced concrete work in the first floor of a housing development. The reinforced concrete area is shown on the key plan on drawing SE/A3.

Column 1

It is unlikely that the location drawings envisaged in F.1.1 (i.e. those drawings required in A.5. 1.a.iii) would identify the reinforced concrete slab. In any event it is better for the taker off to describe the work at taking off stage and then if necessary delete or amend at editing stage.

    Preamble notes for the tests need to be included. The other provisions of F.4.1 will be included in the item descriptions.

Column 2

The side casts for the length first give the dimension from the right hand side of slab as shown on section 5—5 to the outside face of the cavity wall. The bearing of 115 mm is the thickness of the block skin.

    The plan measurement of the slab is diagrammatically shown:

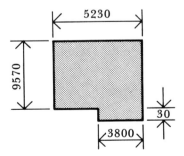

The specification is assumed to be noted elsewhere.

Column 3

The slab is thickened where it bears on walls.

Column 5

The sectional area is in fact 0·03 m². However A.3.4 defines the procedure where limiting dimensions apply and thus this beam is described as not exceeding 0·03 m².

There is no SMM clause for the R.I.W. and it is therefore left to the surveyor's discretion in giving the amount of detail in the description. R.I.W. is in fact a company name which produces a range of products. In this example it can be assumed that the specification is A.1.1, Liquid Asphaltic Composition.

Column 6

The specification for the reinforcement is taken from elsewhere. Reinforcement in attached beams and the like is included in the slab reinforcement and not measured separately (as it was in SMM5).

The bending schedules which are set out in accordance with BS 4466 have the following meanings:

Bar mark	Unique reference to a type and size of reinforcing bar.
Type and size	Prefix letter R = Round mild steel bar Prefix letter Y = High yield high bond round bar Suffix number = Diameter
Number of members	The particular member (e.g. a beam) may be repeated elsewhere. This gives a timesing factor.
Number in each	The number of individual bars in a member.
Total number	Product of previous two columns.
Length	Overall length including allowance for bends, hooks, etc. This is the length entered in the dimensions.
Shape	This refers to Tables 6 and 7 of BS 4466.
A, B, C etc	These are the overall lengths in each plane. Where the shape does not conform to any of the standards listed in Tables 6 or 7 a sketch is given (for example schedule A3/01 for bar mark 8). To avoid omitting any of the bars they should be lined through on the bending schedule when entered on the dimension paper.

Column 9

From information elsewhere the concrete is known to be left exposed and thus fair face is to be measured.

Care must be taken in the measurement of formwork to ensure that adequate preambles for formwork are provided. In this case fair face finish would need to be amplified in a preamble.

Column 10

Since the advent of SMM6, with its encouragement to use bill diagrams, some takers-off have made excessive use of this facility where it is not necessary. In the case of beam 5, measurement of formwork as to edge of slab plus an item for the projection, is quite adequate. No diagram is necessary.

The measurement of the formwork to the edge of the slab is complex because of the varying profiles of edge beams etc. Many of the dimensions have been scaled. To avoid under-measuring lengths of formwork the drawing should first be coloured up showing the different profiles in various colours.

Note the difference between the measurement of the edges of slabs when associated with upstand beams (two items measured) or as in this case, a projecting eaves, when only one item is measured.

Column 11

The formwork to the side of the beam is not strictly in accordance with the SMM. This is because clause F.15.5 requires formwork to be described as being to an item (such as a beam) and stating the number of members in that item. In this way it is possible for the effect of the last sentence of F.15.5. 'Formwork to ends of members shall be deemed to be included with the items' to be priced.

The formwork to the upstand on Section 5—5 is best described by a diagram. Overall dimensions such as those given are adequate for pricing purposes. The formwork to form the throat is measured separately as it will be formed when the slab is laid. There is no SMM clause for this item other than F.13.11.

It is assumed that the tops of the cavity walls which support the slabs will have been closed with a slate or similar and that this will have been measured with the brickwork. There is no need for formwork to the soffit.

Column 12

Although SMM6 clause F.15.5 requires variations in beam profiles to be measured lineally, the variations due to supports on the walls have been enumerated. This is because a total length of 400 mm would cause the item to be enumerated under SMM6 A.7.3.

Column 16

It is assumed that there is no need for a fair finish to the concrete beams. Should a fair finish be required to the soffit or toe of a beam it is suggested that this would best be measured as an extra over item.

Column 17

Landings are deemed to be included in the description of the staircases as F.6.16.

Column 20

Assume the formwork is not fair finish.

Column 21

'Edge of landing' is not strictly in accordance with SMM6. It could alternatively be described as the previous item, 'edge of staircase flight'.

Column 22

The items to the spandril could have been enumerated and the raking cutting included in the description.

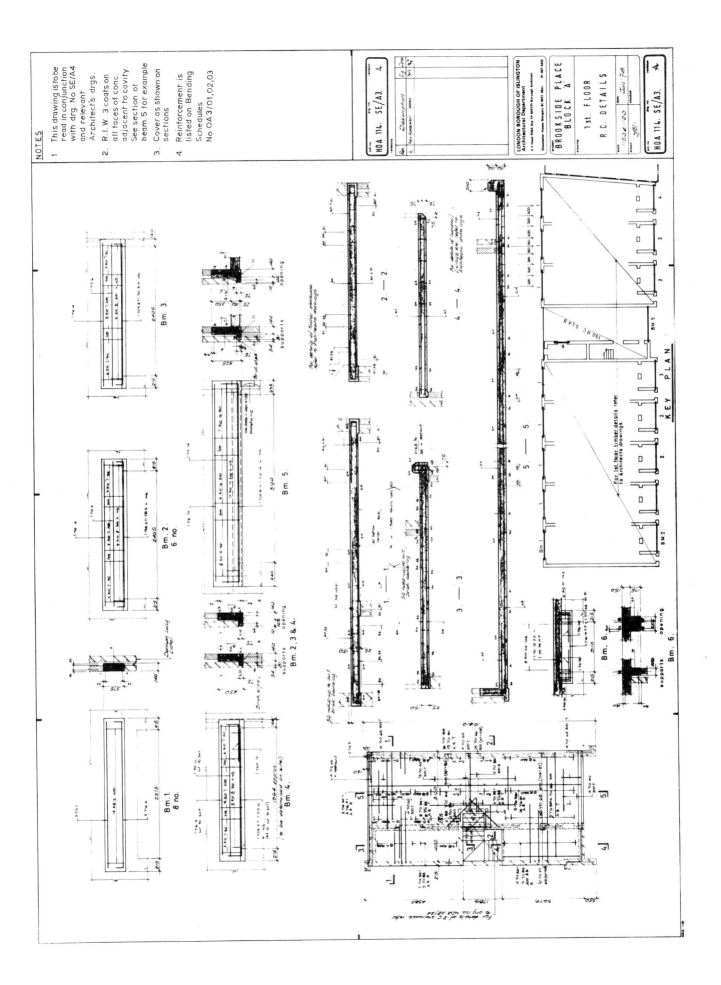

LONDON BOROUGH OF ISLINGTON
Architectural Department

BROOKSIDE PLACE
BLOCK 'A'
1st FLOOR
R.C. DETAILS

HOA 114. SE/A3.

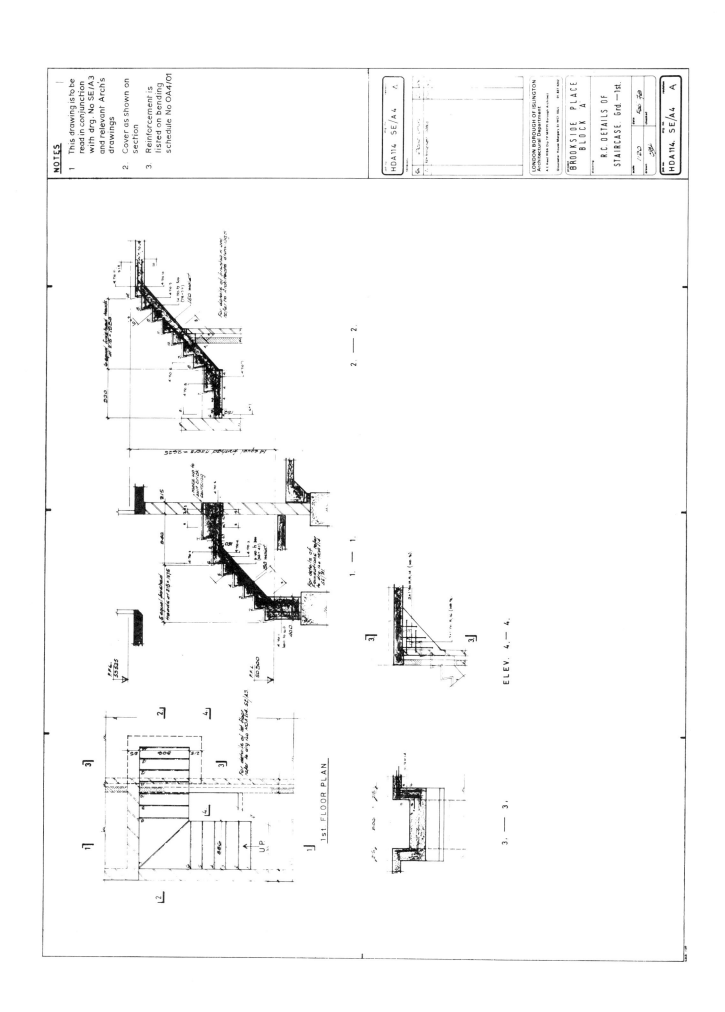

NOTES

1 This drawing is to be read in conjunction with drg. No SE/A3 and relevant Arch's drawings

2 Cover as shown on section

3 Reinforcement is listed on bending schedule No OA4/01

HDA114 SE/A4   A

LONDON BOROUGH OF ISLINGTON
Architectural Department
A J Head RIBA Dip TP MRTPI Borough Architect

Gloucester House Margery St WC1 3QJ   01-837 4242

BROOKSIDE PLACE
BLOCK A

R.C. DETAILS OF
STAIRCASE. Grd. —1st.

scale 1:20

HDA114. SE/A4   A

1st FLOOR PLAN

1. — 1.

2. — 2.

ELEV. 4. — 4.

3. — 3.

Architectural Department

Structural Engineering Group
REINFORCEMENT SCHEDULE
Date

Job							Steel Type Metric R Y X		Drg				Sched.	Rev
BROOKSIDE PLACE									O	A	3	O	I	
No. HDA 114														

Member	Bar Mark	Type & Size	No. of Mbrs	No. in each	Total No.	Length m.m	Shape Code	A m.m.	B m.m.	C m.m.	D m.m.	E/R m.m.
BEAM 1	1	Y12	8	2	16	2700	20	2700				
	2	Y16	8	2	16	3025	38	200	2700			
	3	R8	8	14	112	1000	60	325	90			
BEAM 2 & 3	4	Y16	7	1	7	2650	20	2650				
	5	Y10	7	2	14	2650	20	2650				
	6	Y25	7	1	7	3125	38	300	2650			
	7	R10	7	16	112	1100	81	275	40			
	8	R10	7	9	63	750	99					
BEAM 4	9	Y16	1	1	1	2200	20	2200				
	10	Y10	1	2	2	2200	20	2200				
	11	Y25	1	1	1	2425	37	2125				
	12	Y20	1	1	1	1200	37	900				
	7	R10	1	14	14	1100	81	375	40			
	8	R10	1	8	8	750	99					

150
140
200
140
O/A DIMS

150
140
200
140
O/A DIMS

Member	Bar Mark	Type & Size	No of Mbrs	No in each	Total No.	Length m.m.	Shape Code	A m m	B m.m.	C m.m.	D m m	E/R m.m.
BEAM 5	13	Y16	1	1	1	3700	20	3700				
	14	Y10	1	2	2	3700	20	3700				
	15	Y25	1	1	1	4175	38	300	3700			
	16	R10	1	19	19	1250	81	450	40			
	17	R10	1	12	12	850	99					
BEAM 6	18	Y10	1	2	2	1250	20	1250				
	19	Y16	1	1	1	1575	38	200	1250			
	20	Y16	1	2	2	650	37	500				
	21	R10	1	4	4	825	81	235	40			
	22	R10	1	8	8	1025	60	235	180			

Job: BROOKSIDE PLACE
No. HDA 114

Steel Type Metric RYX

Drg: O A 3 O 2
Sched.
Rev

(shape sketch for bar 17 in column D: 200, 140, 150, 140)

Job	BROOKSIDE PLACE						Steel Type Metric R Y X	Drg.			Sched.		Rev.
No. HDA 114								O	A	3	O	3	

Member	Bar Mark	Type & Size	No of Mbrs	No in each	Total No.	Length m.m.	Shape Code	A m.m.	B m.m.	C m.m.	D m.m.	E/R m.m.
	1	Y10	1	4	4	2100	38	1000	150			
A	2	Y10	1	4	4	2000	62	600	395			
A	3	Y10	1	4	4	1350	62	650	360			
A	4	Y10	1	4	4	1300	62	400	260			
	5	Y10	1	4	4	850	38	400	100			
A	6	Y10	1	4	4	2200	62	650	430			
A	7	Y10	1	4	4	1550	62	800	825			
A	8	Y10	1	4	4	1900	62	900	590			
A	9	Y10	1	4	4	2050	62	400	260			
A	10	Y10	1	4	4	1200	62	400	260			
A	11	Y10	1	4	4	1200	62	500	330			
	12	Y10	1	4	4	1200	54	300	660			
	13	Y10	1	4	4	1050	54	300	500			
	14	Y10	1	4	4	900	54	300	350			
	15	Y10	1	4	4	800	35	600				
	16	Y10	1	4	4	600	35	400				
	17	Y10	1	23	23	800	20	800				

Architectural Department

Structural Engineering Group
REINFORCEMENT SCHEDULE
Date

Job							Steel Type Metric R Y X		Drg				Sched	Rev.
BROOKSIDE PLACE									O	A	4	O	I	A
No. HDA 114														

Member	Bar Mark	Type & Size	No. of Mbrs	No in each	Total No.	Length m m	Shape Code	A m.m.	B m.m	C. m.m.	D m.m	E/R m.m.
SLAB	50	Y16	1	16	16	3100	20	3100				
	51	Y16	1	17	17	3200	34	3100				
	52	Y12	1	11	11	1500	20	1500				
	53	Y12	1	12	12	1600	34	1500				
	54	Y12	1	16	16	2900	38	2550	100			
	55	Y12	1	15	15	3150	38	2800	100			
	56	Y10	1	2	2	2700	20	2700				
	57	Y12	1	3	3	1300	38	950	100			
	58	Y12	1	2	2	1550	38	1200	100			
	59	Y10	1	4	4	1050	38	700	100			
	60	Y12	1	16	16	1500	38	1150	100			
	61	Y12	1	17	17	1250	38	900	100			
	62	Y10	1	21	21	1900	20	1900				
	63	Y10	1	8	8	1250	37	1100				
	64	Y10	1	32	32	4150	20	4150				
	65	Y10	1	4	4	3800	20	3800				
	66	Y10	1	14	14	2250	20	2250				
	67	Y10	1	10	10	3425	20	3425				
	68	Y10	1	20	20	4300	20	4300				
	69	Y12	1	4	4	5125	20	5125				
	70	R8	1	27	27	875	60	225	125			
	71	Y12	1	2	2	1350	38	500	200			
	72	Y12	1	4	23	1800	20	1800				

BROOKSIDE PLACE

BLOCK A

General reinforced
concrete work to load-
bearing brick structure
comprising First Floor
Slab, beams over
openings in external
walls and staircase
from Ground to First
Floor (approximately
.... m³ of in situ concrete
in total volume)

(F.1.1 / F.3.2)

Preamble Notes

1   Tests for materials
2   Tests for finished
      work
(F.4.1.b/c)

Item

Bring to site all
plant necessary
for this section
of the work and
remove on
completion

&

Maintain on site
all plant necessary
for this section

(F.2)

1

First Floor Slab

	Length	Width
	10410	215
		255
Less		255
Return 150		1405
Return 550		3100
Beams 255	955	5230
	9455	
Bearing	115	
	9570	

9·57
5·23
0·15

Reinforced in situ
concrete (28 N/mm²
at 28 days - 20
maximum aggregate)
suspended slab
including beams
100 - 150 thick

(F.6.9)

3·80
0·03
0·15

(5-5
Beam

3·80
0·08
0·08

(toe of
beam

Note - side
notes refer to
detailed plan
and give
position eg
top, left, right
etc

2

109

First Floor Slab (cont)

Bearings

215 wide		255 wide	
Beam 5	3800	Top	1405
Left hand side	4385	Middle	255
	3679	omitting	1400
Staircase	2400	beam	1700
(scaled)	1150	6	4000
	500	Right	
	15914	hand	
Less		side	9570
150			18330
140	290		
	15624	155 wide	
			700
	Left hand corner		1400
			2100

15.62
0.22
0.03

18.33
0.26
0.03

2.10
0.12
0.03

Reinforced concrete
as before suspended
slab 100 - 150 thick

0.70
0.14
0.15

1.62
1.80
0.15

0.72
0.90
0.15

0.91
0.26
0.18

2/ 0.22
0.14
0.18

Ditt

Reinforced concrete
as before suspended
Slab 100 - 150 thick

(Corner

Staircase
1405
215
1620

Beam 6

Add
Ditto

(opening

(supports

3

4

First Floor Slab (cont)

High yield high
bend round
deformed bars to
comply with B S
4449 tied at all
intersections with
1·6 mm diameter
soft annealed
steel wire :

5·23	Reinforced insitu
0·20	concrete (28 N/mm²-
0·15	20) upstand and
	kerb not exceeding
	0·03 m² sectional
	area

(F.G.11)
(Edge of
Slab

(Bar Schedule
OA 3/02
OA 3/03)

3·80	Ditto but 0·03-0·10	16/	3·10	
0·14	m² sectional area	17/	3·20	
0·35			3·70	
	(Beam 5		1·58	
		2/	0·65	

16 Straight and
bent bar in
suspended slab
(F.11.4.c
F.11.5.a)
(1.579 kg/m)

3·80	Apply three coats
0·45	R I W A1.1 liquid
	asphaltic composition
	on sides of concrete
	beam

4·18

25 Ditto
(3.854 kg/m)

16/·11	1·50
12/	1·60
16/	2·90
15/	3·15
3/	1·30
2/	1·55
17/	1·25

12 Ditto

(0.888 kg/m)

Note - Preamble
clause for RIW to
be included

5

6

First Floor Slab (cont)

High yield high bond reinforcement:

4/	5.13
2/	1.35
4/	1.80

12 Straight and bent bar in Suspended slab as before

(0.888 kg/m)

2/	2.70
4/	1.05
21/	1.90
8/	1.25
32/	4.15
4/	3.80
14/	2.25
10/	3.43
20/	4.30
2/	3.70
2/	1.25

10 Ditto

(0.616 kg/m)

7

Mild steel round bar to comply with BS 4499 tied at all intersections with 1.6 mm diameter soft annealed steel wire :

(Bar Schedule (OA 302/ OA 303)

27/	0.88

8 Links and Stirrups in suspended slab
(F. 11.5. c)
(0.395 kg/m)

19/	1.25
12/	0.85
4/	0.83
8/	1.03

10 Ditto

(0.616 kg/m)

8

BP BLOCK A

First Floor slab (cont)

Formwork to
reinforced concrete
to give a fair face
finish :

```
      4385          4385
      1796     215
      3679     255     470
      9860           3915
Less
2/255  510           3679
       9350           255
Less                  3424
projection 150
       9200
```

9.20	To horizontal
3.10	soffite of slab
	(In No 3 separate
3.92	surfaces)
1.41	
3.42	
1.41	(F.15.1.a)

Stairwell

```
              906
    2/215     430
             1336
```

	Ddt
1.34	Ditto
0.70	

9

Formwork to give
a fair face finish:

3.10	To horizontal
	projecting eaves
	of 150 thick
	slab not
	exceeding 250
	deep and with
	projection 81
	wide and 75
	deep
	(In No 1)
	(F.15.5)
	(Beam 5

3.98	To vertical edge (Left
2/1.70	of slab not hand side
0.80	exceeding (Staircase
3.54	250 deep
9.57	(F.15.8) (Left hand side
5.37	(Edges (Right hand side
	(top

3.10	To vertical step
	in soffite of
	slab not
	exceeding
	250 deep
	(F.15.8)
	(Beam 5

10

113

First Floor Slab (cont)

Formwork to give a fair face finish :

2/ 1·41	To one side only Of plain rectangular attached beam 25 deep (In No 17) (F. 15.5)
9·20	(Top & bottom (Right hand side

```
                9 200
Beam 6     910
Staircase 1 800  2 710
                 6 490
```
(centre from column 3 215 bearing (one side only

2/ 6·49	
15·62	
2·10	

5·37	To horizontal upstand 250 wide x 150 deep as Detail 'A' (In No 1)

Detail 'A'
(F. 15.5)

5·37	To throat (In No 1)

11

Formwork to give a fair face finish :

3·10	To plain rectangular horizontal upstand 140 wide x 350 deep (In No 1)

Beam 6

1·34	To plain rectangular attached beam 255 x 175 (In No 1)

2/ 1	Extra over ditto for variation in profile to 140 x 175

1·41	To vertical edge of slab not exceeding 250 deep including 150 wide chamfer on face

(Over Stairwell

12

BP BLOCK A

Beams over openings
in external walls

<table>
<tr><td>8/</td><td>2.80</td><td></td></tr>
<tr><td></td><td>0.38</td><td></td></tr>
<tr><td>6/</td><td>2.84</td><td></td></tr>
<tr><td></td><td>0.38</td><td></td></tr>
<tr><td></td><td>2.76</td><td></td></tr>
<tr><td></td><td>0.38</td><td></td></tr>
<tr><td></td><td>2.32</td><td></td></tr>
<tr><td></td><td>0.38</td><td></td></tr>
</table>

Apply three
coats RIW
to sides of
concrete beam
as before

Reinforced in situ
concrete (28 N/mm²
– 20 aggregate)
isolated beam
0·03 – 0·10 m²
sectional area
        (1n No 16)
        (F. 6.13)

High yield bars
as before

        (Schedule OA3/01)

<table>
<tr><td></td><td>2373</td><td>(B1</td></tr>
<tr><td>2/215</td><td>430</td><td></td></tr>
<tr><td></td><td>2803</td><td></td></tr>
</table>

<table>
<tr><td></td><td>2405</td><td>(B2</td></tr>
<tr><td></td><td>430</td><td></td></tr>
<tr><td></td><td>2835</td><td></td></tr>
</table>

<table>
<tr><td>215</td><td>2405</td><td>(B3</td></tr>
<tr><td>140</td><td>355</td><td></td></tr>
<tr><td></td><td>2760</td><td></td></tr>
</table>

<table>
<tr><td></td><td>1964</td><td>(B4</td></tr>
<tr><td></td><td>355</td><td></td></tr>
<tr><td></td><td>2319</td><td></td></tr>
</table>

(Toes

("

("

Left column takeoff:

<table>
<tr><td>8/</td><td>2.80</td></tr>
<tr><td></td><td>0.14</td></tr>
<tr><td></td><td>0.38</td></tr>
<tr><td>6/</td><td>2.84</td></tr>
<tr><td></td><td>0.14</td></tr>
<tr><td></td><td>0.45</td></tr>
<tr><td></td><td>2.76</td></tr>
<tr><td></td><td>0.14</td></tr>
<tr><td></td><td>0.45</td></tr>
<tr><td></td><td>2.32</td></tr>
<tr><td></td><td>0.14</td></tr>
<tr><td></td><td>0.45</td></tr>
<tr><td>6/</td><td>2.84</td></tr>
<tr><td></td><td>0.10</td></tr>
<tr><td></td><td>0.08</td></tr>
<tr><td></td><td>2.76</td></tr>
<tr><td></td><td>0.10</td></tr>
<tr><td></td><td>0.08</td></tr>
<tr><td></td><td>2.32</td></tr>
<tr><td></td><td>0.10</td></tr>
<tr><td></td><td>0.08</td></tr>
</table>

Right column takeoff:

<table>
<tr><td>14/</td><td>2.65</td></tr>
<tr><td>2/</td><td>2.20</td></tr>
</table>

10 straight and
bent bar in
isolated beam

        (0.616 kg/m)
        (F. 11.4.j)

<table>
<tr><td>16/</td><td>2.70</td></tr>
</table>

12 Ditto
        (0.888 kg/m)

<table>
<tr><td>16/</td><td>3.03</td></tr>
<tr><td>7/</td><td>2.65</td></tr>
<tr><td></td><td>2.20</td></tr>
</table>

16 Ditto
        (1.579 kg/m)

<table>
<tr><td></td><td>1.20</td></tr>
</table>

20 Ditto
        (2.466 kg/m)

<table>
<tr><td>7/</td><td>3.13</td></tr>
<tr><td></td><td>2.43</td></tr>
</table>

25 Ditto
        (3.854 kg/m)

13

14

Beams over openings (cont)

Mild steel round bar as before :

(Schedule OA3/01)

112/	1.00	8 Links and stirrups in isolated beams (0.395 kg/m) (F. 11. 5. c)
112/ 14/	1.10	10 Ditto (0.616 kg/m)
63/ 8/	0.75	

Formwork to reinforced concrete:

8/	2.80	To horizontal isolated beam of plain rectangular section 140 × 375 (In No 8) (F. 17.1)
6/	2.84 2.76 2.30	To horizontal isolated beam 245 × 450 as Detail 'B' (In No 8)

Detail 'B'

BPBLOCK A

Staircase Ground -
First Floor

Reinforced in situ
Concrete (28 N/mm²
– 20 aggregate)
in staircases

(F.6.16)   (Bottom
             step

4/½/ 0.89
      0.40
      0.65

      0.89
      0.28
      0.17

      0.89   (Lower
      1.20   flight
      0.15
                  (Lower
      1.16   waist
      0.91
      0.18
                  (Landing
                   lower
      0.91   940 winder
      0.22   215 –dimensions
      0.03   ———— include
             1155 bearing
                   into wall

½/   1.12   (Wall
      0.91   bearing
      0.18   thickening

                  (Upper
             906   winder
             215
            ————
             1121

6/½/ 0.91   Reinforced concrete
      0.28   in staircases as
      0.17   before

                              (Upper flight
      0.91
      2.06   (Upper waist
      0.15

½/   0.91   Attached
      0.16   beams
      0.16   below
                   waist
½/   1.34
      0.10
      0.10
                   (strings
2/½/ 0.86   as
      0.76   elevation
      0.22   4 - 4

17   18

117

	Staircase Ground - First floor (cont)	

High yield high bond bar :

(Schedule 0A4/01)

4/	2·10
4/	2·00
4/	1·35
4/	1·30
4/	0·85
4/	2·20
4/	1·55
4/	1·90
4/	2·05
3/4/	1·20
4/	1·05
4/	0·90
4·/ 23/	0·80
4/	0·60

10 Straight and bent bar in staircases and strings and associated landings

(0·616 kg/m)

(F.11.4.f)

Formwork to reinforced concrete :

	0·89

To side of upstand beam 450 deep
(In NO 1)
(F. 15.5)
(Bottom step left hand side

	0·89
	0·40

Ditto 570 deep
(In NO 2)
(Bottom step right hand side and end

6/	0·89
1·6/	0·91
	1·30

To undercut risers of staircase not exceeding 250 deep (winder
(F. 15.8)

11

End of riser abutting wall
(F. 15.8)
(top 3 to concrete string

19

20

118

BP BLOCK A

Staircase Ground-First floor (cont)

Formwork to reinforced concrete:

0.92	0.72	To horizontal soffite of solid landing 200 - 300 thick

(In No 1)

(F. 15.1 / F. 15.2)

0.89	1.30
0.91	0.64
0.91	1.06

To soffite of staircase sloping over 15° from horizontal

(In No 3)

(F. 15.1. c)

1.60	0.50

To vertical edge of staircase flight 310 maximum width

(F. 15.8)

0.91

To edge of landing 250 - 500 deep

(F. 15.8)

21

Formwork to reinforced concrete:

0.91

To side of attached beam to landing 25 deep

(In No 1)

&

To side of attached beam to soffite 75 deep

(In No 1)

&

To soffite of ditto 145 wide left - in

(In No 1)

(F. 13.5)

(across cavity on upper flight

2/ ½ /	0.86 0.76

To staircase spandril

2/	1.02

Raking cutting

(F. 13.10)

22

		Staircase Ground-First floor (cont)
2/	1	Formwork to Staircase string 645 + 480 including cutting to ends of three treads and three risers
		(Top three treads
	Item	Protect the inside of the work in this section
		(F.45)
		End of Reinforced Concrete Example
		23

24

EXAMPLE FIVE

Structural steelwork

Project: Warehouse, Reading, Berkshire
Drawing: (PG/81 SF1A)

Generally

A very simple example has been chosen for the illustration of the measurement of steelwork. The example has been contrived to also show the measurement of concrete to a framed structure.

The drawing shows the use of rivets but a more common and economic form of fabrication would be by welding. If welding were used the top cleat would be replaced by a side cleat. In either case, site connections would normally be by bolting.

The drawings are an essential feature for pricing purposes. The steel fabricator calculates the cost of fabrication by reference either to the details of the connections or the performance criteria of the connections (clause P.2.c). The other items of clause P.2.a and P.2.b give information necessary for pricing the erection of the steelwork. The steel members (i.e. beams and columns) are referred on the plan by their serial size (in accordance with BS4 part 1). The actual sizes differ slightly from these serial sizes and it is the actual sizes which determine the overall size of the concrete casings.

Column 2

A description of the work is not necessary as this is provided clearly by drawing SF1A which is also needed to show the information required in clause P.1.2.

Column 4

The requirements of clause P.4.2 are covered by the preamble notes as follows:

P.4.2.a   BS 4360, grade 43
P.4.2.b   BS 4 Part 1
P.4.2.c   By the heading (riveted fabrication, site bolting)
P.4.2.d–f   By the Provisional Sum in column 14. This is necessary because the relevant British Standards do not set levels of acceptability. Testing is expensive and the engineer may decide not to employ any if he is otherwise satisfied with the work. A Provisional Sum is included just in case.

The requirements of clause P.4.3 are covered by the term 'column' and from 'ground floor to roof'.

Column 5

The weight of fittings are aggregated as shown. There is no need to distinguish between fittings to individual columns.

Column 7

There is no need to separate the weight of fittings to first floor and roof level beams.

Column 12

As the beams will be cast at the same time as the floor and roof slabs, they are described as attached beams. The floor is omitted for convenience in this example. There is a difficulty in determining which items should be described as 'in steel framed structures' and as 'other concrete work'. For example, in this case is the slab measured under the heading of steel framed structures? The answer is yes, but no general ruling can be given and it must therefore be at the surveyor's discretion. However, if doubt exists in the surveyor's mind, he should state the method he has chosen with a note in the Bill.

The column is not encased below ground floor level. Each storey height is regarded as being a separate column.

Column 14

As clause F.6.15 regards each storey height as a separate column in the measurement of the concrete, the same principle is applied to the formwork. The 100 mm projection at roof level is not assumed to give rise to a further classification of length of formwork. Had it been, say, a 1·00 m projection, then there would have been three columns. For dimensions in between these examples, it is at the surveyor's discretion and if he is in doubt a note should be included in the Bill as to the method chosen.

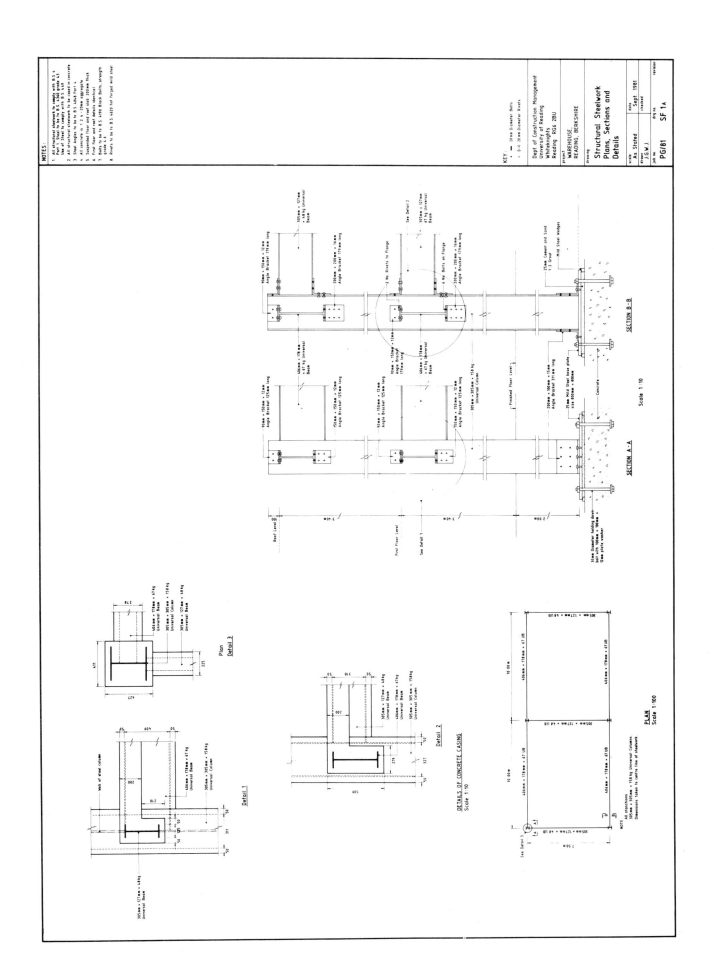

NOTES:

1. All structural steelwork to comply with B.S. 4.
   Part 1 Steel to be to B.S. 4360 grade 43
   Use of Steel to comply with B.S. 449
2. All structural steelwork to be cased in concrete
3. Steel Angles to be to B.S. 4848 Part 4
4. All concrete is 1:2 4:20mm aggregate
5. Suspended floor and roof slab 200mm thick
6. First floor and roof details identical
7. Bolts to be to B.S. 4190 Black Bolts, strength grade 4.6
8. Rivets to be to B.S. 4620 hot forged mild steel

KEY: ● — 20mm Diameter Bolts
     + 0:0 — 20mm Diameter Rivets

Dept of Construction Management
University of Reading
Whiteknights
Reading RG6 2BU

project
WAREHOUSE,
READING, BERKSHIRE

drawing
Structural Steelwork
Plans, Sections and
Details

scale	date
As Stated	Sept 1981
drawn	checked
J.G.W.J.	
job no	drg no	revision
PG/81	SF 1A	

SECTION B - B

SECTION A - A

Scale 1:10

Plan
Detail 3

Detail 1

Detail 2

DETAILS OF CONCRETE CASING
Scale 1:10

PLAN
Scale 1:100

WAREHOUSE READING

STRUCTURAL STEELWORK

Taking-Off List

    Description of work

    Plant

    Bases

    Columns

    Beams

    Erection

    Concrete casing
        (a) plant
        (b) beams
        (c) Columns

    Protection

    Testing

Item	The Contractor is referred to Drawing PG/81 SF1 for a description of the scope of the work and details of connections (P.1.2)
Item	Bring to site and subsequently remove all plant required for this section of the work (P.2.1)
Item	Maintain on site all plant required for this section (P.2.2)

1

2

Bases

6/ 0.80
0.80

25 Cement and sand (1:2) grout under steel stanchion base

(In No 6)

(F.10.1)

6/4/ 1

30 Diameter 150 metric black anchor bolt 350 long including nut and 100 x 100 x 12 plate washer

(F.10.2)

&

Temporary boxing to form 150 x 150 x 300 deep tapered mortice for and including grouting in 350 long anchor bolt in cement mortar (1:2)

(F.10.2)

3

The following in fabricated steelwork with riveted fabrication and bolted connections

Preamble notes

(P.4.2)

Steel to be: to BS4 Part 1

to be Grade 43 (BS 4360)

use of BS 449 angles to BS 4848 Part4

Bolts to be: BS 4190 black bolts grade 46

Rivets to be: BS 4620 hot forged mild steel

2/3400   6800
          2000
           100
          8900

Column

6/ 8.90

Column from ground floor to roof consisting of 305 x 305 x 158 kg Universal Column 8.90 long

(In No 6)

(P.4.3/P.7)

x 158 kg/m

= _____ kg

4

WAREHOUSE READING

__Steelwork - columns__
(cont)

Total from below		Fittings to column (P. 8.1) (Bill after previous item)	6/2/	0·13	150 × 150 × 12 mild steel angle × 27·3 kg/m = _____ kg
6/	0·80 0·80	Steel base plate 25 thick × 196·25 kg/m² = _____ kg	4/2/2/	0·18	200 × 200 × 16 mild steel angle × 48·5 kg/m = _____ kg
6/2/	0·31	100 × 200 × 15 mild steel angle × 33·7 kg/m = _____ kg			__End of fittings to column__

5

6

Beams

3/ 7·17

Beam at first
floor level consisting
of 305 x 127 x 48 kg
Universal Beam
7·17 long
        (In No 3)

    × 48 kg/m
            = _____ kg

            &

Ditto but at roof
level

    × 48 kg/m
            = _____ kg
        (P. 4·3)

4/ 10·00

Beam at first
floor level
consisting of
406 x 178 x 67 kg
Universal Beam
10·00 long
        (In No 4)

    × 67 kg/m
            = _____ kg

            &

Ditto but at
roof level

    × 67 kg/m
            = _____ kg
        (P. 4·3)

6/2/ 0·13

Fittings to beam
consisting of
90 x 150 x 12 mild
steel angle

    × 21·6 kg/m
            = _____ kg
        (P. 8·1)

8/2/ 0·18

Fittings to beam
consisting of
90 x 150 x 12
mild steel angle

    × 21·6 kg/m
            = _____ kg
        (P. 8·1)

7

8

WAREHOUSE   READING

Steelwork - Erection

| | 6/ | 4 | Wedge under stanchion base with steel wedges |

(P. 8. 2. a)

Item   Erect the structural frame, total weight ------ tonnes

(P. 10. 1)

	6/2/	6	75 long M20 150 site black bolts with nut and washer
	2/3/2/	4	(Base
	2/3/2/	2	(305x 127 - 2 floors
	2/4/	6	
	2/4/	2	(402x 178 - 2 floors corners
	2/2/	6	
	2/2/	2	

(P. 8.3)

Note to Worker up

Collate weight as follows

Item	weight

Columns
Fittings to ditto
Beams 1
  "   2
Fittings to ditto
Beams 3
  "   4
Fittings to ditto

× 350·3 kg/1000

= _____ kg

(Ditto but intersection

Total weight _____

9

10

129

<u>Concrete casing</u>

Item	Bring to site and subsequently remove all plant required for the concrete work. (F. 2.1)		

Item

Maintain all plant required for the concrete work.

(F.2.2)

<u>Concrete work to steel framed structures</u>

(F.3.1.b)

$^2/_3/$ 7.07
0.23
0.41

$^2/_4/$ 9.59
0.28
0.51

Concrete (1:2:4 - 20 aggregate) in suspended slabs 150 - 300 thick

(F.6.9)

$^2/_{\frac{1}{2}}/427$ 7 500
427
7073

$^2/_{\frac{1}{2}}/411$ 10000
411
9589

$^2/50$ 409
100
509

$^6/$ 6.90
0.43
0.41

Ditto in isolated casings to steel columns
0.10 - 0.25 m^2 cross sectional area

(In No 12 columns)

(F.6.15)

11

12

130

WAREHOUSE READING

Concrete casing
(cont)

Formwork to
isolated column
casing
427 × 411
(In No 12)

(F.17.2)

2/6/ ½/ 6.90

2/3/ 7.07

Formwork to
horizontal attached
beam casing and
edge of suspended
slab 410 deep one
side × 225 wide
Soffite × 210 deep
other side
(In No 6)

(F.15.5)

Item

Protect all the
work in this
section

2/4/ 9.59

Ditto but 509 deep
one side × 279
Soffite × 309 deep
Other side
(In No 8)

(F.15.5)

Less
509
200
309

Provide the Provisional
Sum of £100 for
testing the steel
Structure

End of Structural
Steelwork Example

13

14

Section Three
Services

Plumbing and mechanical engineering installations

Project: Sotheby Mews Day Centre: Caretaker's house
Drawing: (K 0331 MS 102)
Mechanical engineering specification

Generally

When measuring plumbing installations the sequence is one which follows a pipe from the supply (i.e. mains) through to its ultimate distribution point. This is the best means of avoiding or omitting items which should be measured. The sequence for dealing with individual lengths of pipe is one which measures:

1. The pipe itself.
2. Fittings, connections, etc.
3. Builder's work in connection.

Column 2

The specification included with this example should be amplified and then incorporated into the preambles (see further comment under column 1, example 7).

Column 3

Domestic hot water is a heading required in SMM6. There is no need to define what a pipe contains within such a classification, i.e. no distinction between overflow, cold supply, etc, need to be made. Headings such as rising main are side notes only.

As the extent of the builder's work in connecting the water to the mains is not known a provisional sum is included. An alternative approach would be to measure items such as a trench duct into the building etc. and describe them as provisional. This approach should only really be adopted when the quantities are uncertain but the form of construction is known.

The stop cock is inside the dwelling. A mains stop cock will also be provided by the Water Authority outside the dwelling, usually outside the boundary of the site.

In counting the number of bends on pipes it is often easiest to count the changes in direction in the horizontal plane and then consider the vertical plane.

Column 4

Whether the contractor will use a made bend or an elbow is normally up to him and there is little, if any, difference in cost on small diameter pipes.

Column 5

Having measured the cold supply to the mains tank, the lengths of 1·50 m and 0·60 m are for the sink supply and feed and expansion tank supply.

Column 7

The insulation of the pipe from the combined drain off and stop cock to below ground will be covered by the provisional sum in column 3.

The definition of what is a pipe fitting in clause R.31.4 is stated in the Practice Manual against clause R.10.6.

Column 8

Clause R.38.1 means that builders work should be measured in accordance with the appropriate sections of the Standard Method. The heading in the dimensions is a note for the worker up to make it clear that there is no need to distinguish, for example, between builder's work to the domestic cold water installation and the heating installation.

Although clause V.10.1.a simply states that painting to pipes shall be classified as such, clause V.3.2 defines the qualifications which should be applied as regards size of pipe. As a guide, up to 50 mm diameter pipe is less than 150 mm girth, 100 mm diameter is between 150 mm and 300 mm. Thereafter the measurement of painting is given as an area. The length of pipe painted is the vertical run on both floors and the sink branch.

Making good is where a finish is completed before the pipe is inserted. Labour finishing is where a pipe is inserted and then the finishing takes place. These distinctions can therefore be seen to refer to the sequence of operations in a project. A comment on this is included in Questions and Answers to the sixth edition*.

Column 10

Galvanized cisterns should be painted. There is no specific category in clause V.4 for this type of painting so a full description needs to be given. Flushing cisterns in clause V.10 are cisterns used in connection with WCs etc. and do not apply to storage cisterns.

Column 11

The overflow pipes are not shown on the drawing but are of course necessary and should be measured.

* *Standing Joint Committee for the Standard Method of Measurement (1981), Standard Method of Measurement, 6th edn: Questions and Answers. RICS and NFBTE, London.*

Column 13

The cold water down service is taken down to the first floor level and then the wash hand basin and bath tee in at that level. After supplying the bath the size of the service reduces from 22 to 15 mm.

Column 14

The draincock is at the end of the run so that the pipe can be emptied to facilitate any repairs which may be necessary (such as changing tap washers, etc.).

Column 17

It does not appear necessary (from clause R.33.1) to measure working insulation around the pipes at the connection to the cylinder.

Column 21

It is assumed that the hot water service runs in the floor space to the corner of bathroom. It then surfaces and runs along the wall parallel with the cold water supply.

Column 23

A drain cock is measured for the same reason as column 14. It is not known how many joists will need to be bored through, hence the provisional quantity.

Column 24

It is good practice to include a small sum for the fuel to be used. On a large estate, and especially where central boiler installations are used, the amount of fuel used can be considerable.

Column 25

If cleaning tools are to be provided then they should be described in accordance with clause R.28.

Column 28

The dimension 2/2·60 is taken to be an adequate figure for the vertical rise for cold feed from the feed and expansion tank and the overflow. The length of 1·00 m is to allow for the horizontal pipework around the cylinder.

The flow and return pipes are measured in detail as they do not fall within the classification of header pipes (clause R.10.5). (A header pipe is defined in BS 5643 as 'a manifold connecting a number of pipes, each having a control facility'.) The rise from ground to first floor are assumed to be chased into the blockwork of the chimney breast.

The vertical pipe runs from the ground floor are in a chase and therefore do not need painting.

Column 34

As the lengths of pipes are measured they should be ticked off on the drawing.

The sequence adopted in this case is (directions refer to plan):

In floor space

1. Horizontal right to left (to radiator 4).
2. Horizontal centre to left (to radiator 4).
3. Vertical behind radiator 4.
4. Vertical to radiator 4.
5. Vertical to radiator 6.
6. Horizontal return from radiator 6.
7. Horizontal to radiator 5.

To backgrounds requiring plugging

1. Flow and return drops to radiators 1/2 and 3.
2. Vertical behind radiator 3.
3. Vertical to radiator 3.
4. Horizontal behind radiator 1.
5. Connections between radiators 1 and 2.

The number of bends is difficult to calculate with any precision. This is a generous estimate.

Column 35

The increase of the area by 1/8 is to allow for the additional area over the plan dimensions caused by the panel shape of the radiator and also to allow for the perimeter edges.

Column 38

The time switch is part of the electrical system but a 'To Take' note is a good method of insuring it is not forgotten.

Column 39

The intention of identifying the charge for connecting the gas as a contribution and not a full price is to indicate to the contractor that the scope of the work is greater than indicated by its monetary value.

Sotheby Mews Caretaker's House: Mechanical Engineering Specification

Additional specification notes

Pipework generally to be copper to BS 2871 Part 1 table X with Yorkshire capilliary fittings fixed with standard fixing bands (Yorkshire 101).

Exposed pipework to be painted one coat primer two under coats, one coat oil based paint full gloss finish.

Ball valves to be Portsmouth pattern to BS 1212 with heavy copper float.

Stopcocks	(Yorkshire 508)
Draincocks	(Yorkshire 526LS)
Combine draincock and stopcock	(Yorkshire 525GM)
Gatevalves	(Yorkshire 610)

Insulation to pipes and tanks to be glass fibre rigid sections backed with foil and bound with aluminium self-adhesive tape.

Insulation to cylinder to be purpose made flexible fibre glass mattress backed with 0·35 mm 'Isogenopak' PVC sheet fixed with self-piercing rivets, not less than 8 segments fixed with 3 metal bands.

Inside of galvanized cistern to be painted two coats bituminous paint to BS3416 Part II non-tainting.

Gas fire and back boiler to be Glow Worm 'Capricorn' Model 246. Safety valve to be 15 mm diameter 'Ideal Standard'.

Heating circulating pump to be 'Grundfos Super' reference 4 with union tails for copper pipes.

Radiators to be varying sizes as shown on schedule of 'Stelrad' panel type with 2 No. 15 mm plugs, 1 No. 15 mm air vent, concealed brackets fixed with Duplex Stud anchors reference DX to brickwork. (Note 2 No. loose vent keys to be handed to Caretaker.)

Each radiator to have 1 No. 'Danfoss' thermostatic radiator valve and 1 No. 'Delflo' nickel chrome lockshield head radiator valve.

The Marley Gas Flue System comprises:

1 No. Hearth unit		571 × 444 × 914 overall comprising 1 No. N21, 1 No. N22 and special throating piece.
15 No. Straight blocks	type N3	317 × 117 × 222 high
1 No. Cover block	type N2	368 × 117 × 222 high
4 No. Closer blocks	type N4	260 × 117 × 222 high
8 No. Raking blocks	type N5	365 × 117 × 222 high
2 No. Facing panels	type N15	197 × 51 × 235 with birdguard

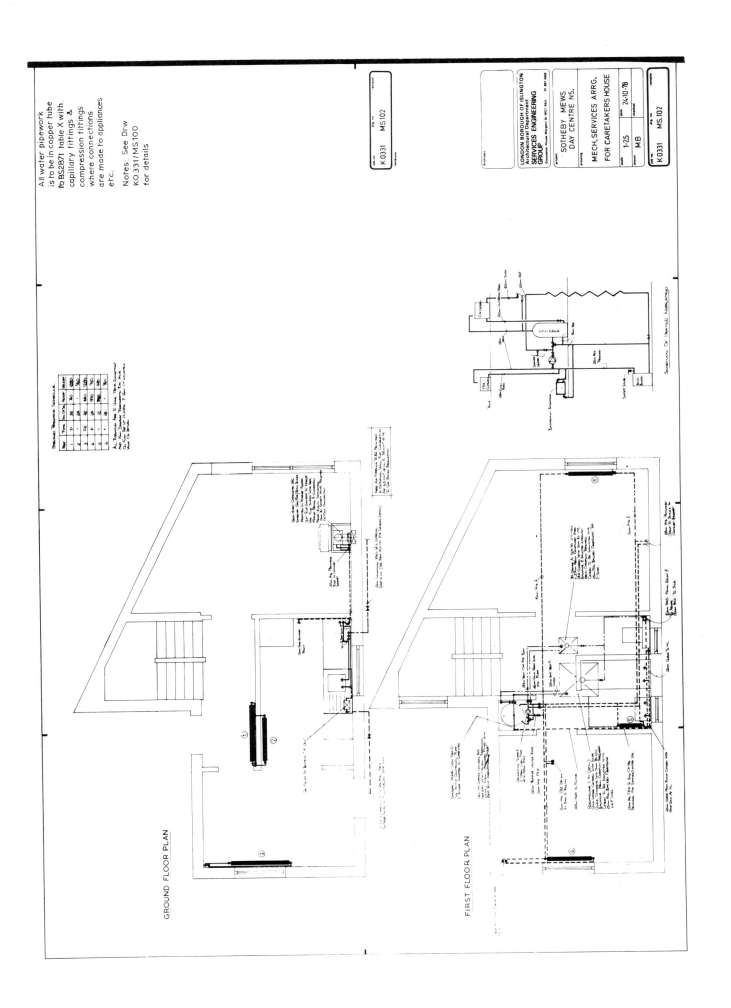

CARETAKER'S HOUSE

MECHANICAL ENGINEERING
INSTALLATIONS

Taking-Off list

Description of work
Drawings and specification
Plant
Domestic cold water

    (a) Connection to main

    (b) Rising main

    (c) Tanks

    (d) Overflows

    (e) Down services

    (f) General items

Domestic hot water

    (a) Cylinder, feed and vent

    (b) Down services

    (c) General items

    (d) Testing

Low pressure heating

    (a) Boiler and flue

    (b) Primary circuit

    (c) Radiator circuit

    (d) General items

Fuel gas

Plans and protection

1

Item	The work in this section comprises the hot & cold water, central heating & gas installation to the Caretaker's House (R.1.1)
Item	The Contractor is referred to the specification for Mechanical Engineering in Preambles and Drawing K0331/MS 102 (R.1.2)
Item	Bring to site and subsequently remove from site all plant required for this section of the work (R. 2)
Item	Maintain all plant for this section of the work (R.2)

2

141

DOMESTIC COLD WATER
(R.4.1.c)

(a) Connection to main

Provide the Provisional Sum of £ 60.00 for connection to water main to be executed complete by the Thames Water Authority (external work)

(R.16)

Provide the Provisional Sum of £ 100.00 for builder's work in connection with water main installation (external work)

(A.2.1)

(b) Rising main

1	15 Combined stop cock and drain cock (Yorkshire 525 GM) and screwed joint to copper pipe
	(R.22)

3

(b) Rising main (cont)

Ground floor	220
	500
	100
	820
Rise 2/2400	4800
Floor space	200
Roof	200
	6020

6·02	15 Copper pipework to BS 2871 Part 1 Table x with 'Yorkshire' capillary fittings and couplers in the running length and fixing with standard fixing bands (Yorkshire 101) at 500 maximum spacing fixed with screws to backgrounds requiring plugging (R.3.6.d/ R10.1/R.15.1)
2·55	15 Ditto but fixed to timber (Roof
	150
	1450
	950
	2550
5	Extra over 15 pipe for made bend (R.13)
1	Extra for 15 equal square tee and capped end (R.10.6)

4

CARETAKER'S HOUSE

DOMESTIC COLD WATER (cont)

(b) Rising main (cont)

2	Extra for 15 equal square tee	
	(R.10.6)	

Sink branch

Vertical	500
Horizontal	900
"	100
	1500

1·50	15 Copper pipework as before
0·60	15 Ditto but fixed to timber (TO F+E tank
2	Extra over 15 (Sink
1	pipe for (Tank made bend
1	Connect 15 (Sink pipe to tap with compression fitting

(R.12.1.b)

5

Tanks

2	15 Stopcock (Yorkshire 508) and compression joint to copper pipe
	(R.22)
	(Connector added back
1	15 'Portsmouth' pattern ball valve to BS 1212 Part 1 with 114 diameter heavy copper float to BS 1968 Class A and joint to copper pipe and fixing to steel tank
	(R.22)
	&
	15 Ditto and fixing to glass fibre tank
2/ 1	Connect 15 service to tank with compression fitting

6

143

DOMESTIC COLD WATER (cont)

(b) Rising main (cont)

Insulation

2·55 0·60	25 Glass fibre rigid section thermal insulation backed with glass reinforced aluminium foil - foil having lapped joints taped with self adhesive aluminium tape mixed around 15 copper pipe. (R.31)
1	Extra for fitting insulation around 15 tee (R.31.4)
2	Ditto around stopcock (R.31.2)

7

Builder's work

NB This is to be billed under a single heading for all installations

(R.38.2)

2/	2·40 1·30	Prepare and paint one coat primer, one undercoat and one coat full gloss finish on copper pipe not exceeding 150 girth — internally (V.10.1.a)
	2	Make good plaster-board around pipe not exceeding 0·30 girth (T.11.2)
	1	Cut and fit 25 softwood board flooring around 15 diameter pipe (N.5.4)

8

CARETAKER'S HOUSE

DOMESTIC COLD WATER
(cont)

(c) Tanks

1	Ferhamglass (C70) glass fibre cold water storage cistern 876 × 775 × 587 of 227 litre (nominal) capacity with loose cover and holes in cistern for 15 ball valve, two 22 down feeds and 28 overflow pipe and hole in cover for 22 vent (R.22)	

&

Galvanized mild Steel cold water storage cistern 610 × 406 × 371 to BS 417 Part 2 (SCM 90) of 54 litre (nominal) capacity with loose cover and holes in cistern for 15 ball valve, 22 down feed and 28 overflow pipe and hole in cover for 22 vent

(R.22)

9

| 1 | 25 Glassfibre insulation as before described to 610 × 406 × 371 Cistern sides and removable top including fitting around four pipes
(R.33.1) |

&

25 Ditto to 914 × 610 × 584 ditto and fitting around five pipes

(For Tank Bearers see Roof Construction)

| 0.64 |
| 0.41 |
| 2.03 |
| 0.37 |

Apply two coats of non-tainting bituminous paint to BS 3416 Part 2 on surfaces of galvanized steel cistern – before fixing

```
 610
 406
1016
 × 2
2032
```

(V.4.1.g)

10

145

DOMESTIC COLD WATER (cont)

(d) Overflows

2/	2.00	28 Copper pipework as before described fixed to timber		

2/ | 1 | Extra over 28 pipe for splay cut end

1 | Connect 28 pipework to steel tank with compression fitting

&

Ditto but joint to glass fibre tank

11

2/ 1 | Hole for pipe through battened, felted and asbestos cement slated roof

(M.7.3)

&

Fix lead slate (supplied by Plumber)

(M.14)

&

Code 5 lead slate 300 x 300 including collar and joint to small copper pipe (fixing measured separately)

(M.54)

12

146

CARETAKER'S HOUSE

DOMESTIC COLD WATER
(cont)

(e) Down services

		From main cistern
	Roof space	700
		1400
		2100

2·10	22 Copper pipework as before described fixed to timber

3·70	22 Ditto and fixing to backgrounds requiring plugging

	Roof	200
	Drop	2400
	Horizontal	1100
		3700

1	Connect 22 pipework to steel tank with compression fitting & 22 Gatevalve (Yorkshire 610) and joints to copper pipe

3	Extra over 22 pipe for made bend

13

2	Extra for 22 reducing sweep tee (Basin Bath

2/ 1·00	15 Copper pipework as before described to backgrounds requiring plugging (Basin Bath WC
1·70	

	700
	1000
	1700

2/ 1	Extra over 15 pipe for made bend
2	

2	Connect 15 pipework to tap with compression fitting (WC/WWP Bath/Basin

1	Ditto but to water waste preventer

1	15 Draincock (Yorkshire 526 LS) and screwed joint to copper pipe

14

147

DOMESTIC COLD WATER
(cont)

(e) Down services (cont)

2·10		25 Glass fibre insulation as before to 22 pipe
2/	3·70 1·00 1·70	Prepare and paint two coats as before on copper pipe not exceeding 150 girth — internally
	1	Make good plaster-board around pipe not exceeding 0·30 girth & Hole through 102 brick wall for small pipe and make good (G.49.1)
	2	Make good wall plaster around pipe not exceeding 0·30 girth

1		Cut and fit 3 thick coated hardboard bath panel around 22 diameter pipe

(f) General items

Item		Mark the position of holes, mortices, chases and the like in the structure for the Domestic Cold Water Installation. (R.37.2)
Item		Allow for chlorination of cold water installation as described in the Preambles
Item		Allow for testing the cold water installation as described in the Preambles (R.37.5)

DOMESTIC HOT WATER

(R.4.1.e)

(a) Cylinder, feed and vent

1	400 Diameter × 1050 high indirect hot water storage cylinder (IMI Range Ltd Reference B 2 c) with a nominal capacity of 114 litres and integral thermostatic valve and by-pass
	&
	75 Thick thermal insulation to sides and top of 400 diameter × 1050 cylinder comprising purpose made flexible glass fibre mattress backed with 0.35mm 'Isogenopak' PVC sheeting fixed with self piercing rivets and securing with three metal bands

17

1·75 / 2·69	22 Copper pipework fixed to timber

Roof
Feed 1300
250
200
1750

roof space 200
tank 587
overflow 2/100 200
horizontal run } 500 1200
2687

2·70 / 1·35	22 Ditto but fixed to background requiring plugging

Drop Feed 2400
Return 300
2700

Vent
floor/ceiling 2400
Less cylinder 1050
1·350

1	Connect 22 pipework to galvanized steel tank with compression fitting
2/1	Ditto but to copper cylinder

18

149

DOMESTIC HOT WATER (cont)

(a) Cylinder, feed and
      vent (cont)

1.80		Prepare and paint two coats paint on copper pipe not exceeding 150 girth — *internally*
2.60		
2.00		

2/ 4	Extra over 22 pipe for made bend	

2		Make good plaster- board around pipe not exceeding 0.30 m girth

1	22 Gate valve (Yorkshire 610) and joints to copper pipe	

2.69	25 Thick glass fibre insulation as before to 22 pipe	

19

20

DOMESTIC HOT WATER
(cont)

(b) Down services

From vent

1	Extra over 22 pipe for equal sweep tee

Drop to floor

400
1200
1600

1·60 1·40	22 Copper pipework as before fixed to backgrounds requiring plugging

In floor

3·10	22 Ditto but fixed to timber

21

4	Extra over 22 pipe for made bend
2	Ditto for reducing sweep tee (Basin Bath
2/ 1·00 1·20 2·60 1·80	15 Copper hot water service as before with standard bands fixed to backgrounds requiring plugging

Suk   Drop
1·500   2400
200   200
100
1800   2600

7	Extra over 15 pipe for made bends
4	Connect 15 pipework to tap with compression fitting

22

DOMESTIC HOT WATER (cont)

(b) Down services (cont)

1	15 Draincock (Yorkshire 526 LS) and screwed joint to copper pipe

1.2/	1.60
	1.40
	1.00
	1.20
	2.60
	1.80

Prepare and paint two coats of paint on small copper pipe not exceeding 150 girth – internally

2	Cut and fit 25 softwood flooring around 22 pipe (N.5.4)

1	Ditto around 15 pipe

8	Bore 38 softwood floor joist for 22 pipe (Provisional) (N.29.1)

23

1	Make good plasterboard around pipe not exceeding 0.30 m girth

(c) General items

Item	Mark the position of holes, mortices, chases and the like in the structure for the Domestic Hot Water Installation (R.37.2)

Item	Allow for testing the domestic hot water installation as described in the Preambles (R.37.5)

Provide the Provisional Sum of £10.00 for gas to be used in connection with the testing of the hot water and heating installations (R.37.5)

24

152

CARETAKER'S HOUSE

LOW PRESSURE HEATING INSTALLATION

(R.4.1.e)

(a) Boiler and flue

1	Glow-worm 'Capricorn' Model No 246 combined gas fire and back boiler as described in the Preambles and fixing in position in hearth unit (measured separately) and connecting to gas flue system (R.22.1) & 15 Safety valve (Ideal Standard B1) jointed to copper pipe NB Vents for Gas Regulations already measured with 'External Walls'

25

'Typex 150' gas flue system (Marley Buildings Ltd - True Flue Division) bedded, jointed and pointed in 'Flue Joint' ready mixed mortar in accordance with the manufacturers instructions and building into blockwork leaf of cavity walling

(See Typex Catalogue)

1	Special recess panel 610 high to suit 'Glow worm' 'Capricorn' boiler in three pieces (Reference RB/RSL and RSR) & Special lintel to match last (Reference RL) & Flue back Offset block 190 x 190 x 222 (Reference c150) (R.40.4/G.54)

26

LOW PRESSURE HEATING (cont)

(a) Boiler and flue (cont)

'Typex 150' (cont)

15	Single flue block with bonding nib 266 × 152 × 222 (Reference 5×1)
8	Raking flue block 260 × 152 × 222 (Reference 4×11)
1	Single flue cap unit 572 × 508 × 76 (Reference 9×9)
2	38 Concrete louvre 190 × 300 (Reference ChX)

End of 'Typex 150'

0.60 / 0.60	Prepare and apply two coats RIW on surfaces of concrete flue vertically over 225 wide
.15 / 8 / 0.27 / 0.22	(in cavity (G.37.2)

27

(b) Primary circuit

2	Connect 22 pipework to boiler with screwed joint

Primary Flow and Return, Cold Feed and Vent

2 / 2.60 / 1.00	22 Copper pipework as before fixed to backgrounds requiring plugging

2 400
200
2600

2 / 2.60	22 Ditto fixed to backgrounds requiring plugging fixed in chase

Primary Vent and Flow and Return

1.2 / 2.20	22 Ditto fixed to timber backgrounds
2 / 3.80	(laid in floorspace

28

CARETAKER'S HOUSE

LOW PRESSURE HEATING
                    (cont)

(b) Primary Circuit (cont)

$\dfrac{5}{3}$		
2	Extra over 22 pipe for made bend	(Flow (Return (Cold feed
2/ 1	Ditto for equal sweep tee	(Cold feed & vent (Flow to Cylinder
2	Ditto for reducing sweep tee	(Cylinder
1		(Return for radiator circuit
1	Connect 22 pipework to steel tank with compression fitting	

To Cylinders

2/ 0.50	15 Copper heating service	(No bands or fixing
2	Extra over 15 pipe for made bend & Connect 15 pipework to copper cylinder with screwed joint	
1	'Grundfos super' (Reference 4) two speed circulating pump as described in the Preambles with 22 union tails and joint to copper pipe	

29

30

155

LOW PRESSURE HEATING
(cont)

(b) Primary circuit (cont)

2/ 2·60 1·00 0·50	Prepare and paint two coats of paint on copper pipe not exceeding 150 girth – internally
2	Make good plaster- board around pipe not exceeding 0·30m girth & Hole through 102 brickwall for small pipe and make good (Floor space & Cut and fit 25 Softwood flooring around 22 diameter pipe
2/ 8	Bore 38 softwood floor joist for 22 pipe (Provisional)
2·60	Cut chase in blockwork for two 22 diameter pipes and make good

31

(c) Radiators

'Stelrad' panel type ready
primed radiators with
two 15 plugs and one
15 air vent with
concealed brackets fixed
with 'Duplex' stud
anchors (Reference Dx)
to brickwork including
mortice :

See Schedule
on Drawing

1	Single panel 480 × 740 high (P5 & Single panel 960 × 590 high (P4
2	Single panel (P2 960 × 740 high 6
1	Double panel 1280 × 440 high (P3 & Single panel 1280 × 740 high (P1

32

LOW PRESSURE HEATING (cont)

(c) Radiators (cont)

(Bathroom radiator and main circuit

2	Loose vent keys for 'Stelrad' for 15 radiator air vents (R.28)	

6	15 'Danfoss' thermostatic radiator valve with built in sensor and joints to radiator and Copper pipe	

&

15 'Delflo' (2388) lockshield head radiator valve with nickel chrome finish and joints to radiator and Copper pipe

3/ 1 — Extra over 22 pipe for reducing sweep tee

8.90	15 Copper pipework with bands fixed to timber background
3.80	
2.00	
1.00	
1.20	
1.60	
2/ 0.60	

(In floor space

Radiator
6
2.20
Less 1.00
1.20

4/ 2.60	15 Ditto fixed to background requiring plugging
2.00	
0.60	
1.50	
2/ 0.40	

51 — Extra over 15 pipe for made bend

33

34

LOW PRESSURE HEATING (cont)

(c) Radiators (cont)

8	Extra over 15 pipe for equal sweep tee	

1	22 Gatevalve (Yorkshire 610) and joints to copper pipe & 15 Drawcock (Yorkshire 526 LS) and screwed joint to copper pipe	

Painting radiators

0.48 0.74 0.96 0.59 2/ 0.96 0.74 1.28 0.44 1.28 0.74	Spot prime, paint one undercoat and one coat full gloss finish to surfaces of primed metal radiators – internally (v.7) $+\frac{1}{8}$ = _____ m²	

(Panel section

35

4/	2.60	Prime and paint two coats paint on copper pipe not exceeding 150 girth – internally

1	Hole through 102 brick wall for small pipe and make good	
2		

(In floor space

2/	2	Make good plasterboard around pipe not exceeding 0.30 m girth

36

158

				CARETAKER'S HOUSE

LOW PRESSURE HEATING (cont)

(c) Radiators (cont)

	4		Make good wall plaster around pipe not exceeding 0.30 girth (Radiators 1 & 2

3/2/	1		Cut and fit 25 softwood flooring around 15 diameter pipe (Radiators 6 5 4

	16		Bore 38 softwood floor joist for 15 pipe (Provisional)

37

(d) General items

Item		Mark the position of holes, mortices, chases and the like in the structure for the low pressure heating installation (R.37.2)

Item		Allow for providing a circuit control diagram, valve number schedule and for labelling valves and for providing a schedule of spares (R.37.6)

Item		Allow for testing the heating installation as described in the Preambles (R.37.5)

To take Sangramo time Switch and electrical supply with electrical installation

38

FUEL GAS

(R.4.1.g)

Provide the Provisional Sum of £65.00 for the contribution charge for gas carcassing and connection to gas main to be executed complete by the North Thames Gas Board

(R.16)

(NB. this is a contribution charge not the full price of the work

Builders'work

1 | Hole through 265 brick cavity wall for small pipe and make good

&

Hole through 150 reinforced concrete bed for ditto

---

1 | Hole through 102 brick wall for small pipe and make good

1.20
1.40 | Prime and paint two coats of paint on copper pipe not exceeding 150 diameter – internally

GENERALLY

Item | Allow for preparing all drawings and working instructions required for the mechanical installations as described in the Preambles

(R.37.6)

Item | Protect the whole of the work in this section

(R.41)

End of Plumbing and Mechanical Engineering Installation

EXAMPLE SEVEN

Electrical installations

Project: 463–471 Liverpool Road
Drawings: Electrical specification
(HDA 140/ES 104)
Master electrical symbols

Generally

This is the measurement to the electrical installation in a four person dwelling on a housing estate. It would be as well to measure the installation to each dwelling separately and any timesing for other dwellings carried out on the summary of the Bill of Quantities.

A section of the dwelling has not been provided here as all that it is necessary to learn from such a drawing is the floor to ceiling height which is 2·40 m.

Specification for electrical accessories and fittings

Item	Manufacturer	Reference No.	Notes
1. Consumer control unit	J.A. Crabtree	List No.206/1	Six way – 3/15 Amp, 1/30 Amp, 1/40 Amp 1 Blanking off plate – List No.191/1
2. Trunking	BS 4678 Part 1 Class 2		75 x 75 mm Single Compartment
3. Switched socket outlet	M.K. Electric Ltd	List No.2535 WH1	Steel box List No.86621C One Gang Type
4. Ditto	M.K. Electric Ltd	List No.2536 WH1	Steel box List No.88621C Two Gang Type
5. Cooker cable outlet unit	M.K. Electric Ltd	List No.5044 WH1	Steel box List No.87721C
6. Cooker switch	M.K. Electric Ltd	List No.5215/CK/WH1	Steel box List No.878WH1
7. Batten holder light fitting	Rock Electrical Accessories	List No.BH313P	
8. Pendant fitting 300 mm drop	Rock Electrical Accessories	List No.CR310	Lampholder List No.73W 1·0mm^2 3 core flexible cord to BS 6500 Table 9
9. Plate switch for lights	M.K. Electric Ltd	List No.4870 WH1	Steel box List No.86621C One Gang-one way
10. Ditto	M.K. Electric Ltd	List No.4871 WH1	Steel box List No.86621C One Gang-two way
11. Ditto	M.K. Electric Ltd	List No.4872 WH1	Steel box List No.86621C Two Gang-two way
12. Ceiling switch for lights	M.K. Electric Ltd	List No.3191 WH1	Bathroom light switch
13. Time switch for boiler	AMF Venner Limited	'Venotime'	Include black enamelled steel circular conduit box
14. Boiler switch (double pole)	M.K. Electric Ltd	List No.5423 WH1	Steel box List No.86621C
15. Telephone cord outlet plate	M.K. Electric Ltd	List No.3541 WH1	Steel box List No.86621C
16. TV Coaxial socket outlet	M.K. Electric Ltd	List No.3520 WH1	Steel box List No.86621C

161

Column 1

Clause S.1.2 of SMM6 states that a detailed specification of the electrical system together with drawings indicating the scope of the work shall be provided. There is no further explanation of what scale of drawing for example is needed to show the scope of work. In this project it would probably be appropriate to use a smaller scale drawing (ES 104 is very detailed) provided the specification information included on ES 104 is included in the specification. Each scheme needs careful consideration and the quantity surveyor should resist providing the estimator with every piece of information he has just because it is available.

An area of concern to the authors is the reference to Specification' in clause S.1.2. As the specification is not a contract document what reliance can a quantity surveyor place on the contractor agreeing to abide by it? There seems no reason why the specification should not be incorporated into the preambles to the Bills of Quantities and this problem is then resolved.

The specification for electrical accessories and fittings included with this example needs to be amplified by a specification covering cables, conduits, etc.

The division of work into internal and external is new to SMM6 and is an important one to note.

Column 2

A detailed discussion on the appropriateness of a provisional sum for work of this nature is given in Example 8, Column 35. A 5·00 m length of trench is taken from the estate layout drawings.

Column 3

It is necessary to lay ducts for electrical cables with open joints so that any water which may enter the duct can drain away. The draw wires need to be measured as they need to be installed by the main contractor for use by others and do not fall under the heading of SMM clause S.16.1. The ducts are measured in accordance with clause W.6.1 via clause S.26.1.

Column 5

The heading 'Mains installation excluding final subcircuits' must be included in the Bill as must all subsequent similar headings. The consumer control unit has 3 no. 15 amp breakers for the 2 lighting and 1 boiler circuits, 1 no. 30 amp breaker for the power circuit and 1 no. 40 amp breaker for the cooker circuit. The blanking off plate is a spare point for any later circuits which may need to be installed.

The fixing of equipment and accessories and the need to describe the nature of the background are given in clause S.3.6.

The trunking is from the ground floor slab to the consumer control unit (c.c.u.) to cover up the incoming mains. The trunking must be properly connected to the c.c.u. to maintain the earthing.

The meter tails are the connection from the meter to the c.c.u.

Column 6

A distribution sheet is an essential aid to measurement and is also required by clause S.21.1. If one is not provided one should be prepared as below. (A useful further guide is by Watson, R., 'SMM6 The measurement of final subcircuits', Chartered Quantity Surveyor, Vol.2, Nos. 1 and 2, pp 13—15.)

Distribution sheet

Location	Lighting installation			Power installation		Remarks
	Circuit	Fittings	Switches	Circuits	Outlets	
Ground Floor						
Lobby	1	1	1 x 1 way, 1 gang	—	—	
Hall	1	1	1 x 2 way, 2 gang 1 x 2 way, 1 gang	3	1 single	
Lounge	1	1	1 x 1 way, 1 gang	3	2 single, 1 double	TV outlet Telephone outlet
Dining/Kitchen	1	2	2 x 1 way, 1 gang	3	1 single, 3 double	
Dining/Kitchen	—	—	—	4	1 cooker unit 1 cooker switch	
Dining/Kitchen	—	—	—	5	1 time switch 1 boiler outlet 1 connection to boiler	
Cupboard	1	1 (b.h)	1 x 1 way, 1 gang	—	—	
TOTAL	Circuit 1	6 points	5 x 1 way, 2 x 2 way			
First Floor						
Store	2	1	1 x 1 way, 1 gang	3	1 single	
Landing	2	1	1 x 2 way, 1 gang	3	1 single	
Toilet	2	1	1 x 1 way, 1 gang	—	—	
Bathroom	2	1 (b.h)	1 x 1 way, pull cord	—	—	
Bedroom 1	2	1	1 x 1 way, 1 gang	3	1 single, 1 double	
Bedroom 2	2	1	1 x 1 way, 1 gang	3	1 single, 1 double	
TOTAL	Circuit 2	6 points	5 x 1 way, 1 x 2 way	Circuit 3	8 single 6 double	

Comments on distribution sheet

The two lighting circuits are distinguished as between ground and first floor and they also need to be measured separately because of the different types of fixings of cables.

The power circuit is a single circuit to both floors.

There is a separate circuit for both boiler, with its associated time switch, boiler socket and connection to boiler. It is perhaps worth raising a query as to whether a room thermostat should be added. One of the problems in measuring subcircuits as complete units occurs when variations are made.

Column 7

The numbers of socket outlets is taken from the distribution sheet. The nature of the background could have been described in the remarks column or a further column could have been added for this information. The laminated partitions would probably come under the classification S.3.6.c but a query should be raised with the architect.

Column 9

It is instinctive to want to describe the different backgrounds to which the conduit and cable are fixed. However in *Questions and Answers** it states that final subcircuits do not need to be classified according to the surfaces on which they are fixed because the layout is at the discretion of the contractor.

The difference between the types of cables to the power circuit is that the conduit acts as the earth to the cooker installation.

Column 18

A major item in any electrical installation is the earthing system. It is not clear from the SMM when general earth bonding and earthing each installation should be given as an item (under clause S.25.2) and when it should be measured in detail under clause S.22. In view of the global approach taken to the measurement of final subcircuits, it would seem reasonable that earthing is dealt with in the same way (i.e. clause S.25.2). However, because this issue is not clear the earthing is measured in detail in this example and a general item for completing the earthing and bonding is also measured on a 'belt and braces' approach. If it were to be measured as an item it is suggested that the number of points at which the cable is earthed are described and the number of clamps at each point stated.

The lengths of cable are calculated on the basis of clause S.17.1.a, i.e. 0·30 m added at c.c.u. and at earth point.

This earth from the sink is assumed to run horizontally in a conduit on the surface of the wall into the soil and ventpipe duct, rise vertically and then in a conduit as far as the bath. An earth wire not in a conduit runs from the basin and connects both to the bath and the twin socket in Bedroom 1.

Column 20

Clause S.24 deals with identification plates of some financial significance.

Although this item is not measured in accordance with SMM6 it is a reasonable approach to the low value labels etc. associated with an installation of this kind.

---

* *Standing Joint Committee for the Standard Method of Measurement (1981), Standard Method of Measurement, 6th edn: Questions and Answers. RICS and NFBTE, London.*

The noggins are for the necessary fixing of ceiling lighting points. (One of the 12 lighting points is fixed to a partition.)

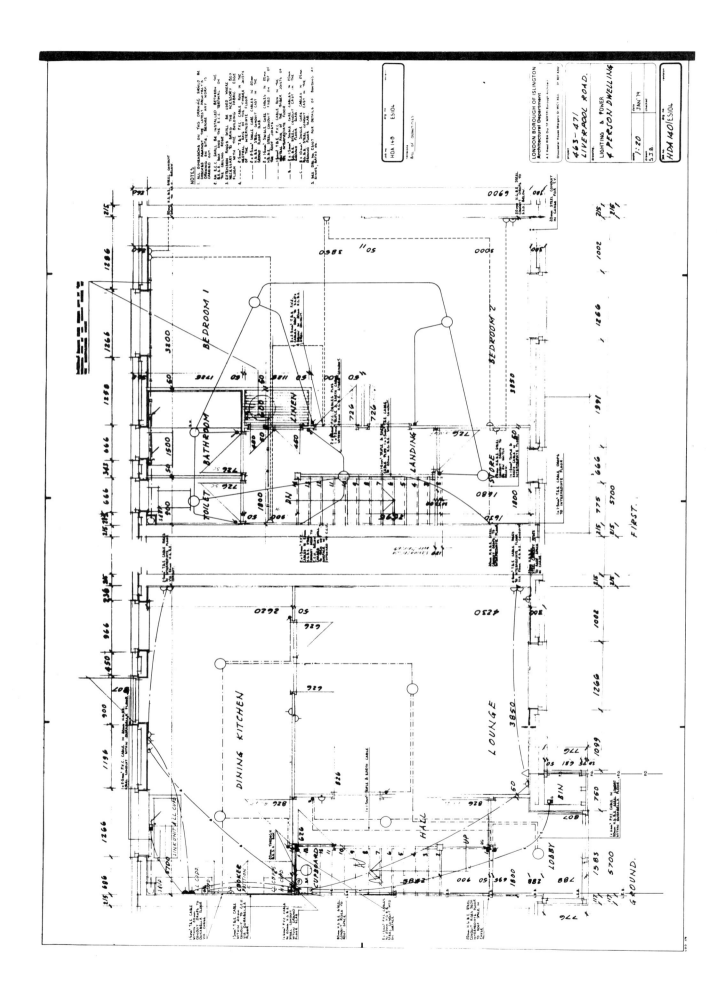

LONDON BOROUGH OF ISLINGTON
Architectural Department

463-471
LIVERPOOL ROAD.

LIGHTING & POWER
4 PERSON DWELLING

1:20

S.J.B.

JAN '74

HDA 140/ES104

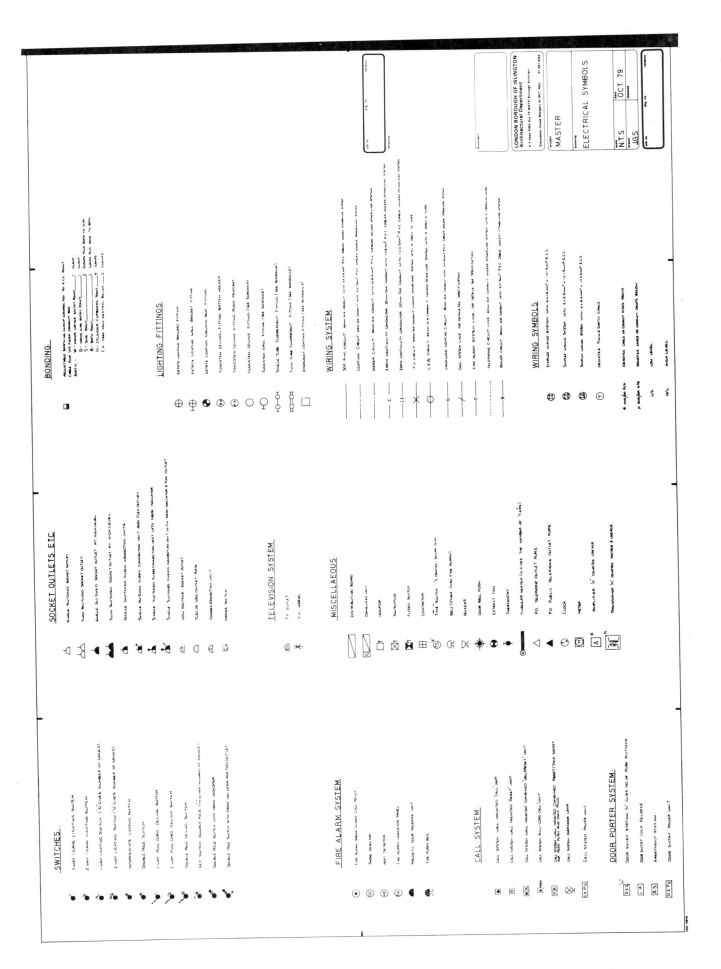

LIVERPOOL ROAD

ELECTRICAL INSTALLATION

for

Four person house

Taking-off list

1 Description of the work
2 Plant
3 Connection to supply
     Provisional Sum
     Trench
     Ducts
4 Mains installation
   excluding final sub-circuits
5 Power installation
6 Lighting installation
7 Electrical work associated
   with plumbing and
   mechanical engineering
   installations
8 Telephone and TV installations
9 Earthing system installation
10 General items and builders'
    work

GENERALLY

The Electrical Installations
are as shown on Drawing
HDA 140/ES 104 and the
Electrical Specification in
the Preambles

          (S.1.2)

All work in this section
is internal unless
otherwise described

          (S.5)

1

PLANT

Item.   Allow for bringing
        to site and
        subsequently
        removing from
        site all plant
        required for this
        section of the work

          (S.2)

Item   Allow for maintaining
        on site all plant
        required for this
        section of the work

          (S.2)

CONNECTION TO SUPPLY

Provide the Provisional
Sum of £110.00 for
provision of electricity
supply executed
complete by the
Electricity Board

          (S.20)

2

169

ELECTRICAL INSTALLATION
(cont)

CONNECTION TO SUPPLY (cont)

Builders work
in connection
(S. 26.2)

| | 1 | Extra over 100 Clayware duct for easy bend |

&

Hole through 265 hollow wall for large pipe and make good
(G.49)

&

Hole through 150 reinforced concrete bed for large pipe
(F.9.9)

5.00 — Excavate trench to receive pipe not exceeding 200 nominal size starting at ground level not exceeding 2 000 deep and average 750 deep including earthwork support, grading bottom, backfilling and compacting with selected material and removing surplus spoil from site (External work)
(S.27.1)

5.20 — 100 Diameter Clayware duct laid with open joints
(S.26.1)

&

Draw wire for Electricity Board's main

5.00 — 100 Diameter Clayware duct laid with open joints (External work)
(S.26.1)

&

Draw wire for Electricity Board's main (External work)

3

4

170

	LIVERPOOL ROAD	
	ELECTRICAL INSTALLATION (cont)	
	MAINS INSTALLATION EXCLUDING FINAL SUB-CIRCUITS (S.4.1.c)	
1	Consumer control unit for surface mounting (J A Crabtree and Company Limited List No 206/1) with three 15 amp breakers, one 30 amp breaker and one blanking off plate (Crabtree List No 191/1) fixing with screws to background requiring plugging (S.6.2)	
1.50	75 x 75 Stove enamelled steel cable trunking surface type to BS 4678 Part 1 Class 2 in single compartment with connecting sleeves in the running length and cable retaining clips fixing with screws to background requiring plugging (S.12)	

75x75 Trunking

4	Extra for blank end (S.12.2)
	&
	Connection to equipment or control gear with flanged couplings (S.12.3)

Meter tails

3.00	16 mm^2 Single Core 300/500 Volt PVC insulated and PVC sheathed copper conductor cable to BS 6004 Table 4 in meter tails and drawing into trunking (S.17.1)

5

6

POWER INSTALLATION

(S. 4. 1. d)

3	Flush one-gang switched socket outlet (MK Electric Limited List No 2535 WHI) with steel box (List No 866 21C) and fixing with screws to background requiring plugging (S.8.1/s.10)	
5	Ditto and fixing to laminated partitions (S.8.1/s.10)	
6	Flush two-gang switched socket outlet (MK Electric Limited List No 2536 WHI) with steel box (List No 886 21C) and fixing with screws to background requiring plugging (S.8.1/s.10)	
1		Flush cable outlet unit for electric cooker (MK Electric Limited List No 5044 WHI) with steel box (List No 877 21C) and fixing with screws to background requiring plugging (S. 8.1) & Flush fixing cooker switch (MK Electric Limited List No 5215/CK/WHI) with steel box (List No 878 21C) and fixing ditto (S. 8.1)

7

8

LIVERPOOL ROAD

ELECTRICAL INSTALLATION

POWER INSTALLATION (cont)

LIGHTING INSTALLATION

(S.4.1.e)

1	Final socket Outlet sub-circuit installation consisting of $2.5\,mm^2$ PVC triple and earth cable installations and two $2.5\,mm^2$ single core cable in 20 diameter heavy galvanized black enamelled (h g b e) conduit with fourteen switched socket outlet points
	&
	Final cooker outlet sub-circuit installation of two $10\,mm^2$ cable in 25 h g b e Conduit with one cooker outlet point and one switch

2	Surface fixing batten holder (Rock Electrical Accessories Limited List No BH 313 P) with black enamelled steel circular conduit box and extension ring and fixing with screws to timber
10	Surface fixing pendant fitting with 300 drop comprising ceiling rose (Rock Electrical Accessories Limited List No CR 310) lampholder (List No 73 W) and $1.0\,mm^2$ three core flexible cord to BS 6500 Table 9 and black enamelled steel conduit box and fixing with screws to timber

9

10

LIGHTING INSTALLATION
(cont)

1	Toilet Flush one gang one way plate switch (MK Electric Limited List No 4870 WHI) with steel box (List No 866 ZIC) and fixing with screws to background requiring plugging	3	Flush one gang two way plate switch (MK Electric Limited List No 4871 WHI) with steel box (List No 866 ZIC) and steel backing plate and fixing to laminated partitions
7	Ditto with steel backing plate and fixing to laminated partitions	1	Flush two gang two way plate switch (MK Electric Limited List No 4872 WHI) with steel box (List No 886 ZIC) and steel backing plate and fixing to laminated partitions
		1	Surface fixing one way ceiling switch (MK Electric Limited List No 3191 WHI) and fixing with screws to timber (Bathroom

11

12

LIVERPOOL ROAD

ELECTRICAL INSTALLATION

LIGHTING INSTALLATION
(cont)

<table>
<tr><td>1</td><td>Final lighting sub-circuit installation consisting of 1.5 mm² PVC triple and earth cable and two 1.5 mm² single core cable in 20 hg be conduit to first floor comprising six lighting points, five one way and one two way switch points

&

Ditto but to ground floor comprising seven lighting points, five one way and two two way switch points</td></tr>
</table>

13

ELECTRICAL WORK
ASSOCIATED WITH PLUMBING
AND MECHANICAL
ENGINEERING INSTALLATIONS

(S.4.1.h)

<table>
<tr><td>1</td><td>'Venotime' surface fixing time switch (AMF International Limited) with black enamelled steel circular conduit box and fixing with screws to backgrounds requiring plugging

&

Flush fixing double pole switch (MK Electric Limited List No 5423 WHI) and steel box (List No 866 21C) and fixing ditto

&

Final boiler sub circuit installation consisting of two 1.5 mm² single core cable and 20 diameter hg be conduit installation to ground floor with one boiler outlet point, one time switch point and connecting one boiler</td></tr>
</table>

14

175

TELEPHONE INSTALLATION

(S.4.1.j)

1	Flush fixing telephone cord outlet plate (MK Electric Limited List No 3541 WHI) with steel box (List No 86621C) and fixing with screws to background requiring plugging	(Lounge
Item	The telephone installation will be executed by British Telecom free of charge and the Contractor is to allow here for providing general attendance	(B.12)

15

5.00	Excavate trench for pipe not exceeding 200 nominal size as before (External work)
6.00	100 Diameter clayware duct as before (External work)
1	Extra for easy bend as before & Hole through 265 hollow wall for large pipe as before & Hole through 150 reinforced concrete bed for large pipe as before

16

176

LIVERPOOL ROAD

ELECTRICAL INSTALLATION

TELEVISION INSTALLATION

(S.4.1.r)

1	Flush fixing TV coaxial socket outlet (MK Electric Limited List No 3520 WHI) with steel box (List No 866 21C) and fixing with screws to background requiring plugging
5.55	20 Heavy gauge screwed welded steel black enamelled conduit to BS 4568 Parts 1 and 2 Class 2 complete with fittings and draw wire and fixing with crampets in chases in brickwork and blockwork

Floor / ceiling 2/2.40 4800
Floor thickness   250
      5050
Roof space   500
      5550

17   **177**

EARTHING SYSTEM INSTALLATION

(S.4.1.r)

	5.40 / 1.40	20 Black enamelled heavy gauge screwed welded steel conduit embedded in concrete (S.11.1.d) (Gas main
	1.20 / 2.40	20 Ditto but to surfaces with screws to backgrounds requiring plugging (S.11.1.a) (Sink
	2.00	20 Ditto but fixing to timber surfaces (in first floor
2/	5.40 / 0.30	6mm² Single core PVC green insulated cable drawn into conduit (S.17.2) (Gas main
5/	4.00 / 2.30 / 0.30	2.5 mm² Single core PVC green insulated cable fixed to timber (Basin/Bath/cylinder 1.900 400 2300
	1.40 / 1.20 / 2.00 / 2.40	2.5 mm² Ditto (Sink drawn into conduit (Bath (Sink
2/2/	0.30	

18

EARTHING (cont)

5	Flush fixed cord outlet plate and box (Crabtree List No 2075) and fixing with screws to backgrounds requiring plugging

2/	1	Adjustable earthing clamp (Tenby Electrical Accessories Limited List No EC15) and fixing to small pipe (Gas (Sink (Basin (Bath (Cylinder
2/	1	
2/	1	
4/	1	

	1	Connect 2.5 mm² cable to sanitary fitting or other similar appliance (Sink (Bath
	1	

1	Connect Electricity Board's earth wire to Consumer Control unit earth lug

Item	Complete the earthing and bonding in accordance with IEE Regulations

19

GENERALLY

Item	Mark the positions of holes, mortices, chases and the like for the electrical installations

(S.25.1)

&

Allow for the provision of all identification labels, charts etc as required by the Preambles

(NOT SMM)

&

Allow for testing & the provision of a test certificate

(S.25.5)

&

Allow for the provision of drawings and working instructions as described in the Preambles

(S.25.6)

20

LIVERPOOL ROAD

ELECTRICAL INSTALLATION

BUILDER'S WORK

(S.27.6)

12	Cutting away and making good for concealed h g b e conduits to lighting points	
14	Ditto for socket outlet point	
1	Ditto for fitting outlet point	(Cooker
3		(Boiler
1		(TV
1		(Telephone
1	Cutting away and making good for exposed conduit, switches and outlets to control gear point	
		(Consumer control unit

11	50x50 sawn softwood noggin 450 long (Provisional)	
	Protection	
Item	Allow for protecting the work in this section (S.28)	

End of four person house

End of electrical example

EXAMPLE EIGHT

Drainage

Project:	Stanmore Street	
Drawings:	Drainage layout	(HDA 154 01/2 A)
	Manhole and gully schedule	(HDA 154 01/3)

Generally

This is the drainage to a new housing development. Although the drawings and schedules indicate 12 manholes, manhole No.12 will be undertaken on a separate contract and is therefore not included in this measurement. A manhole schedule is provided. Where none is provided the taker-off should prepare such a schedule and include it in his taking-off.

Column 2

The re-establishment of the water table level (D.3.1(a) via W.1.3) may well be at a different time and give a different value from that for the general excavation.

Column 3

In the measurement of drainage there is the problem of deciding from which level to commence excavation of manholes and drain runs. It is normal for the site first to be excavated to a reduced level. Therefore the manholes etc can be dug as from the cover level which is assumed to equate with the reduced level. A check of existing ground levels should always be made to see if there is any making up of levels. If there is, a reduction would have to be made in the depth of excavation. The concrete bed is 150 mm thick and 100 mm is added for the benching and pipe thickness. The 250 mm is then added to depth to give depth of excavation. Specification 80 (published by the Architectural Press) recommends that for manholes over 1800 mm deep that the base should be 225 mm thick and this should be checked with the architect.

Column 6

Next to roadway includes next to footpath and in this case, manholes 5, 6 and 7 are approximately 600 mm from the edge of the footpath. Clause D.19 requires this qualification when the depth of excavation is greater than the horizontal distance.

Column 7

Small manholes do not need a concrete slab on top of the manhole as the cover can sit directly on the brickwork. The height of the brickwall is therefore calculated as depth of excavation less 150 mm concrete bed and less 50 mm cover.

Column 8

For the large manholes a cover slab is necessary. The brickwork height is therefore the excavation depth, less 150 mm bed less 150 mm slab less 50 mm cover.

Column 9

The fair face is the inside girth of the manhole times the height of brickwork less the benching in the bottom. The oversailing course is needed to support the manhole covers on the small manholes (797 x 797 mm) only. The drawings show oversailing courses on all manholes but this should be taken as being diagrammatic only. The note on the schedule against 'Construction' namely, 'Concrete slabs on top where applicable', overrides the drawing.

The fair face to margins covers the soffits of the two course oversailing brickwork.

Column 10

Slabs of this nature would almost certainly be precast.

Column 11

The sequence of measuring building in the 100 mm pipes is as follows:

1	manhole 8
1.1.1.2/3	manholes 2, 4, 5, 6, 10
1.1.2/2	manholes 1, 3, 7, 11

The order shown above is not unique but is the result of carefully lining through the schedule as the items are entered on the dimension paper.

Column 12

None of the 100 mm diameter channels need to be the full length of the manhole.

Columns 13 and 14

Takers-off should always check references to B.S. manufacturers catalogues and the like to ensure that references are up to date. If necessary, the items can be entered in the dimensions as shown on the drawings, but a note should be made to ensure that current catalogues are ordered if none are available. Then these will have arrived by the time the Bills are ready for editing. The NAT references are to the Standard Illustrated catalogue of the Clay Pipe Development Association, dated November 1979. The Broads references on the drawings are no longer correct as Broads (now renamed Sandell Perkins and Broads) do not issue a catalogue. The manhole covers are supplied by Drainage Castings (Burn Bros). References are altered accordingly and this is a case in support of the need to check catalogue references and refer to the Architect.

Step irons are provided in manholes; those numbered 1, 3, 5 and 7 are classified as inspection chambers on the schedule and therefore step irons are not necessary.

Column 15

A clear concise schedule is essential for measuring drain runs. The main runs should be taken first. It is assumed that the reduced surface level is constant between both ends of each drain run. This will almost certainly always be the case and the rounding to the nearest 0·25 as required by W.3.1. indicates that precision is not needed.

Where drain runs pass under the footpath/service vehicle access they have been surrounded in concrete as shown on the detail for drain sections under roads (runs 1–2, 2–3, 8–9, 9–10). The length of drain runs are scaled as the length between manholes. It is not considered necessary to make precise adjustment for the fact that the excavation and pipe lengths differ slightly due to the pipe passing through the manhole wall and that the pipe is also laid to a gradient and is therefore longer than the horizontal length.

In an *Introduction to the Sixth Edition* of SMM6* the author states that where a trench moves from one 2 m depth increment to another it is not necessary to divide the total length into its separate stages. It is felt that this is a practical solution to the measurement of drain trenches and one which is realistic in terms of the way the work will be done on site. However, in *Questions and Answers*† it states that although clause W.3.1. is similar to D.11, nevertheless drain branches continue to be banded within depth ranges. This pamphlet† is not wholly consistent in that against clause D.11 it does suggest that surveyors should use flexibility in itemizing the work to suit the circumstances of the site. In this respect if may be suitable to make a qualification to SMM6 along the lines proposed by Goodacre*. For the purpose of illustrating how trench langths are banded, the method proposed by the Standing Joint Committee† is the one used in this example.

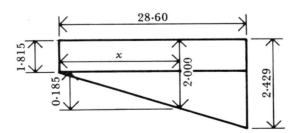

* *Goodacre, P.E. (1979), Standard Method of Measurement; An Introduction to the 6th edn. College of Estate Management, Reading.*

† *Standing Joint Committee for the Standard Method of Measurement (1981), Standard Method of Measurement, 6th edn: Questions and Answers. RICS and NFBTE, London.*

MH4—MH9

Total length	28.60
Depth range	1·815
	2·429

By proportion:

$$2 \cdot 000 \quad less \quad \underline{1 \cdot 815} \qquad\qquad 2 \cdot 429 \quad less \quad \underline{1 \cdot 815}$$
$$0 \cdot 185 \qquad\qquad\qquad\qquad\qquad 0 \cdot 614$$

$$\frac{x}{0 \cdot 185} = \frac{28 \cdot 60}{0 \cdot 614}$$

therefore
$$x = \frac{28 \cdot 60 \times 0 \cdot 185}{0 \cdot 614}$$

$$= 8 \cdot 60$$

MH8—MH9

By proportion:

$$2 \cdot 000 \qquad\qquad\qquad\qquad 2 \cdot 429$$
$$\underline{1 \cdot 817} \qquad\qquad\qquad\qquad \underline{1 \cdot 817}$$
$$0 \cdot 183 \qquad\qquad\qquad\qquad 0 \cdot 612$$

therefore
$$x = \frac{14 \cdot 00 \times 0 \cdot 183}{0 \cdot 612}$$

$$= 4 \cdot 20$$

Columns 23 and 24

The lengths shown in the schedule are scaled direct from the drawings or are subtotals as calculated below.

Location	Length		Total
1	2/4·50 + 3·00	=	12·00
2	13/1·00	=	13·00
3	6/5·00	=	30·00
4	6/5·00	=	30·00
7	4/2·50	=	10·00
9	5/4·00	=	20·00
10	5/1·00	=	5·00
11	2/2·00	=	4·00
12	9·00 + 7·00 + 3·00	=	19·00
23	2/2·00	=	4·00
25	2·00 + 3·00	=	5·00

These subtotals would be shown on the schedule if space permitted. As branch runs are entered on the schedule they should be lined through on the drawing. The maximum depths of trenches are taken from manhole invert levels or from an approximation of the depths of the various runs. No greater accuracy is necessary. Two bends per branch have been measured. This represents a suitable figure at taking-off stage. When the work is executed on site the bends in particular should be remeasured. The old rule of thumb 'One bend per branch plus two bends per gully plus three bends per fresh air inlet' has tended in the authors' experience to be overgenerous.

Column 25

Drain runs 1, 2, 3, 4, 5, 6, 7, 8 and 16 can all be said to fall within the range not exceeding 1·00 m deep. However, drain run 9 is calculated as follows and it can be seen that the calculations of the excavation for branch drains into 'banded' depth stages (see comment above under Column 15) becomes excessively tedious.

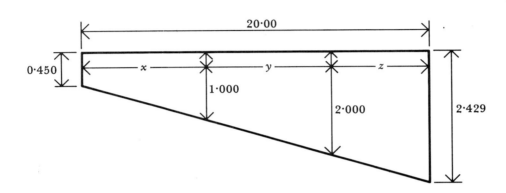

$$\frac{x}{(1 \cdot 000 - 0 \cdot 450)} = \frac{x + y}{(2 \cdot 000 - 0 \cdot 450)} = \frac{20 \cdot 00}{(2 \cdot 429 - 0 \cdot 450)}$$

$$x = \frac{20 \cdot 00 \times 0 \cdot 550}{1 \cdot 979}$$

$$= 5 \cdot 60$$

$$x + y = \frac{20 \cdot 00 \times 1 \cdot 550}{1 \cdot 979}$$

$$= 15 \cdot 70$$

Therefore $\quad y = 15 \cdot 70 - 5 \cdot 60$

$$= 10 \cdot 10$$

and $\quad z = 20 \cdot 00 - 15 \cdot 70$

$$= 4 \cdot 30$$

By similar proportion for drain runs 10–12

Drain run 10

$$
\begin{aligned}
x &= 1 \cdot 40 \\
y &= 2 \cdot 50 \\
z &= \underline{1 \cdot 10} \\
& \underline{\underline{5 \cdot 00}}
\end{aligned}
$$

Drain run 11

$$
\begin{aligned}
x &= 1 \cdot 10 \\
y &= 2 \cdot 00 \\
z &= \underline{0 \cdot 90} \\
& \underline{\underline{4 \cdot 00}}
\end{aligned}
$$

Drain run 12

$$
\begin{aligned}
x &= 5 \cdot 30 \\
y &= 9 \cdot 60 \\
z &= \underline{4 \cdot 10} \\
& \underline{\underline{19 \cdot 00}}
\end{aligned}
$$

New depth calculation for drain run 13:

$$
\begin{aligned}
& 1 \cdot 615 \\
less & \underline{0 \cdot 450} \\
& \underline{1 \cdot 165}
\end{aligned}
$$

$$
\begin{aligned}
x &= \frac{13 \cdot 00 \times 0 \cdot 550}{1 \cdot 165} \\
&= 6 \cdot 10 \\
y &= 13 \cdot 00 - 6 \cdot 10 \\
&= 6 \cdot 90
\end{aligned}
$$

By similar proportion for drain runs 15, 16A, 21

Drain run 15

$$
\begin{aligned}
x &= 7 \cdot 00 \\
y &= \underline{8 \cdot 00} \\
& \underline{\underline{15 \cdot 00}}
\end{aligned}
$$

Drain run 16A

$$x = 2.80$$
$$y = \underline{3.20}$$
$$\underline{\underline{6.00}}$$

Drain run 21

$$x = 7.30$$
$$y = \underline{8.20}$$
$$\underline{\underline{15.50}}$$

New depth calculation for drain run 18

(Note: take as all not exceeding 2.00 in depth)

$$2.115$$
less $$\underline{0.600}$$
$$\underline{\underline{1.515}}$$

$$x = 2.10$$
$$y = 8.00 - 2.10$$

$$= 5.90$$

By similar proportion for drain run 20

$$x = 3.30$$
$$y = \underline{9.20}$$
$$\underline{\underline{12.50}}$$

New depth calculation for drain run 22

$$1.500$$
less $$\underline{0.450}$$
$$\underline{\underline{1.050}}$$

$$x = 1.90$$
$$y = 4.50 - 1.90$$

$$= 2.60$$

By similar proportion for drain run 25

$$x = 2.60$$
$$y = \underline{2.40}$$
$$\underline{\underline{5.00}}$$

New depth calculation for drain run 23

		2·000			2·429
less		1·250	less		1·250
		0·750			1·179

$$x = 2.50$$
$$y = 4.00 - 2.50$$
$$= 1.50$$

Drain run 26

Similar to run 23

$$x = 4.50 - 2.50$$
$$= 2.00$$

Drain runs 24 and 27

Similar to run 9

Run 24

$$x = 3.90$$
$$y = 7.10$$
$$z = 3.00$$
$$14.00$$

Run 27

$$x = 2.50$$
$$y = 4.50$$
$$z = 2.00$$
$$9.00$$

Column 28

By inspection of the drawings where pipes pass under roads etc. adjustments to provide a concrete surround as opposed to benching are made. The drawings should be coloured up to indicate the lengths where such adjustments are necessary and then lined through on the drawings when entered on the dimension paper.

Column 32

Although the gully schedule distinguishes between gullies which have side or back inlets the taker-off should realize that the gullies will be the same price and can therefore be grouped together. There is a tendency to include in Bills of Quantities information which is not necessary for pricing purposes and this trend should be carefully controlled.

Column 34

The NAT references are substituted for the Broads references.

Column 35

In the *Questions and Answers* published by the Standing Committee* an issue is raised over the use of 'Provisional Sums'. Clause W.8 requires the connection to sewers, which can only be undertaken by a Local Authority, to be measured as a Provisional Sum. Clause A.8.1.a however states that Provisional sums should only be used for work which 'cannot entirely be foreseen, etc'. The connection to the sewer can of course be quite clearly foreseen, but in their answer the Standing Joint Committee says that a Prime Cost Sum should only be used where a statutory undertaking will fulfil all the obligations of a nominated subcontractor. Clause B.11.1 refers to an authority which does not accept the terms of nomination. Clause 6.3 of the 1980 J.C.T. Standard Form of Contract also states that the provisions of the contract do not apply 'to the execution of part of the Works by a Local Authority'.

Columns 37 and 38

It is good practice to carefully check the drawings to see that all the items included have been transferred onto the dimension paper. In this example it was noticed that soil and vent pipes are to have access plates at ground floor level. It is quite possible that this information will not be on the drawings used for measuring the internal plumbing so a 'To Take' note is a useful reminder for whoever measures that section of work.

It is not normal for takers-off to write preambles in detail but certain items which appear on the drawing should be entered on the dimensions. In this way the correct alternatives (if such exist) in standard offices preambles (perhaps even national preambles such as the National Building Specification) are identified or new clauses highlighted which will need to be written in full.

Column 38

The item for this fresh air inlet looks inelegant but for such items where many small items combine to form a functional item it is best to describe in a single item. Any estimator will know what is required and will price accordingly. It is never wrong to use plain English!

* *Standing Joint Committee for the Standard Method of Measurement (1981), Standard Method of Measurement, 6th edn: Questions and Answers. RICS and NFBTE, London.*

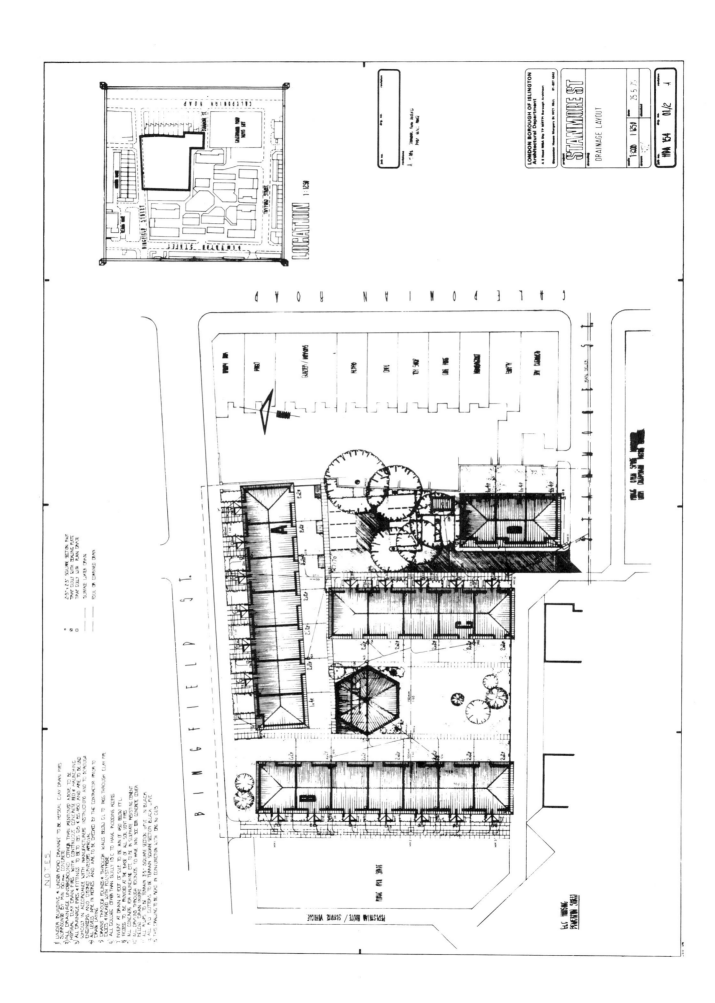

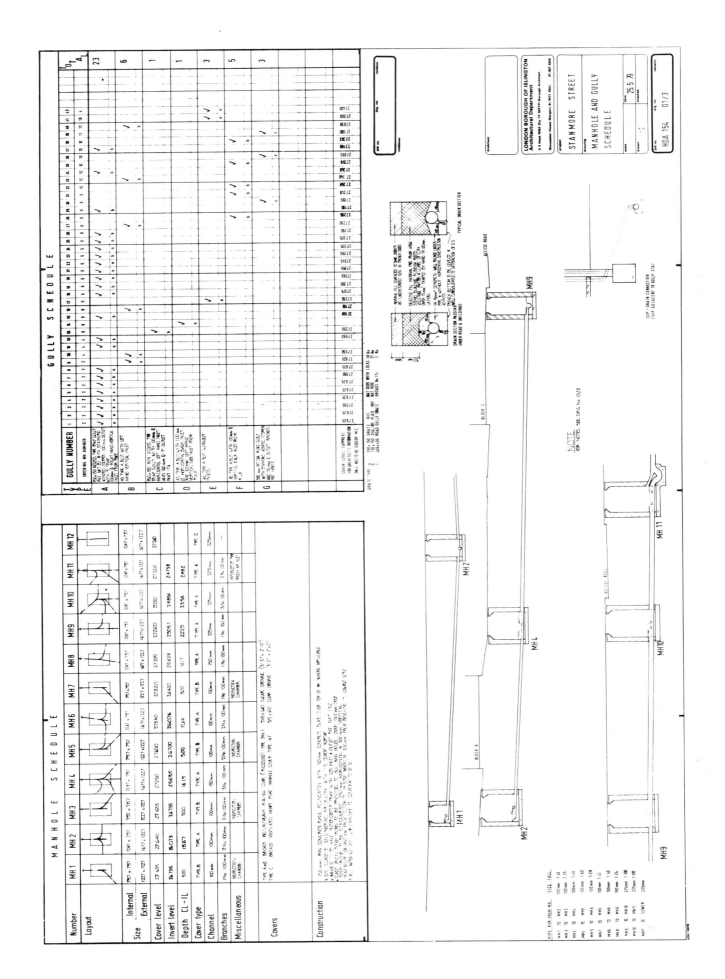

STANMORE STREET

DRAINAGE

Taking-off list

Nature of ground and
        ground water level

Plant

Manholes
      Excavation
      Earthwork support
      Bases
      Walls
      Cover slab and
         corbels
      Benching
      Holes for pipes
      Channels and
         branches
      Interceptor
      Covers
      Step irons
      Keys

Main runs (Schedule)

Branches (Schedule)

Gullies

Connection to sewer

General items

The drainage layout and
details are shown on
Drawings 01/2A and 01/3
        (W.1.1)

The Contractor is referred
to the trial hole information
given in the Preambles to
'Excavation and Earthwork'
for the nature of the ground
        (W.1.3)

Ground water was found
at 23.00 above sea level
on 10 April 1981
        (W.1.3)

Item   Bring to site and
        subsequently remove
        from site all plant
        required for this
        section of the work
        (W.2.1)

Item   Maintain all plant
        for this section of
        the work
        (W.2.2)

2

THE FOLLOWING IN
NO 11 MANHOLES

(W.7.1)

NB Cover level
    is taken as
    Reduced Level

(a) Large manholes
    not exceeding 2.00 m

MH	Depth CL-IL	1677 × 1227 Base + pipe thickness	Total depth
2	1567	250	1817
4	1615	250	1865
6	1264	250	1514
8	1617	250	1867
		Total	7063

(b) Large manholes 2.00 –
    4.00 m

9	2229	250	2479
10	2394	250	2644
11	2482	250	2732
		Total	7855

(c) Small manholes

1227 × 1227

1	900	250	1150
3	900	250	1150
5	900	250	1150
7	900	250	1150
		Total	4600

1.68
1.23
7.06

Excavate pit
commencing at
reduced level not
exceeding 2.00 m
maximum depth
     (In No 4)
     (D. 13.5)

&

Remove surplus
excavated material
from site
          (D. 29)

1.68
1.23
7.86

Excavate pit
commencing at
reduced level not
exceeding 4.00 m
maximum depth
     (In No 3)
     (D. 13.5)

&

Remove surplus
excavated material
from site

          (D. 29)

3

4

194

STANMORE STREET

No 11 | MANHOLES (cont)

<table>
<tr><td>1·23
1·23
<u>4·60</u></td><td>Excavate pit not exceeding 1·25 m on plan in both directions commencing at reduced level not exceeding 2·00 m maximum depth
(In No 4)
(D·13·5)</td></tr>
</table>

&

Remove surplus excavated material as before

<u>Earthwork support</u>

MH 1,2,3,4,8   ne 2m

MH 5,6,7   ne 2m next roadway

MH 9,10,11   ne 4 m

Large MH	Small MH
1227 1677 <u>2904</u> × 2 <u>5808</u>	1227 1227 <u>2454</u> × 2 <u>4908</u>

5

2/ | 4·91
1·15 | Earthwork support not exceeding 2·00 m maximum depth with faces not exceeding 2·00 m apart

| 5·81
1·82

2/ | 5·81
1·87 | (D.15/D.17)

2/ | 4·91
1·15
5·81
1·51 | Ditto <u>next</u> roadway

(D.19)

5·81
<u>7·86</u> | Earthwork support not exceeding 4·00 m maximum depth with faces not exceeding 2·00 m apart

<u>Item</u> | Keep the surface of the site and the excavations free of surface water
(D.25)

c

No 11 | MANHOLES (cont)

(b) Large manholes

MH	Excavation depth	Base and Cover	Wall height
2	1817	350	1467
4	1865	350	1515
6	1514	350	1164
8	1867	350	1517
9	2479	350	2129
10	2644	350	2294
11	2732	350	2382
		Total	12468

Bases

4/	1·23
	1·23
	0·15
7/	1·68
	1·23
	0·15

150 Plain sulphate resisting cement concrete (14 N/mm² – 20 aggregate) bed spread and levelled over 100 but not exceeding 150 thick

(F. 6.8)

Girth (a)

1227 × 4 = 4908

Less 4/225   900

                        4008

Girth (b)

1227 + 1677 = 2904
                        × 2
                        5808

Less 4/225   900

                        4908

4/	1·23
	1·23
7/	1·68
	1·23

Level and compact bottom of excavation

(D. 40)

Walls

(a) Small manholes

MH	Excavation depth	Base and cover	Wall height
1	1150	200	950
3	1150	200	950
5	1150	200	950
7	1150	200	950
		Total	3800

4·01
3·80
4·91
12·47

One brick wall in engineering bricks Class B (48.5 N/mm²) in English bond in sulphate resisting cement mortar (1:3)

(G.5.3.a)

7

8

STANMORE STREET

No 11 MANHOLES (cont)

	Fairface		
Girth	4008	4908	
Less 4/225	900	900	
	3108	4008	
Height		3800	12467
Less 4/200	800		
Less 7/200			1400
	3000	11067	

3·11
3·00
4·01
11·07

Extra over engineering brickwork for fair face and flush pointing as the work proceeds

(G. 14.3)

4/ 3·11

Oversailing engineering brickwork two courses high and 102 total projection

(G. 5.4)

4/2/ 3·11

Extra over engineering brickwork for fair face to margins

(G.14.6)

9

Covers to large manholes

7

150 Precast sulphate resisting cement reinforced concrete (21 N/mm² – 20 aggregate) cover slab 1677 × 1227 reinforced with square mesh reinforcement to BS Reference A 142 weighing 2·22 kg/m² with rebated opening for 915 × 610 manhole cover and frame (measured separately) and bedded in cement mortar (1:3) on brick sides to manhole

(F.18)

4

Plain concrete (14 N/mm² – 20 aggregate) benching in bottom of manhole 797 × 797 finished to steep slopes and average 225 high with edges rounded and dished to main channel and branches and trowelled smooth

(W.7.5)

10

197

No 11 | MANHOLES (cont)

7	Benching as before 1247 × 797

Holes

1	Build in end of 100 drain pipe to one brick wall in engineering brickwork in sulphate resisting cement mortar (1:3) and make good fair face and flush pointing one side (W.7.4)
2/ 1.1.2/ 3	
1	
1.1.2/ 2	
1	

2/ 2	Ditto 150 drain pipe ditto
1	

3/ 2	Ditto 225 drain pipe ditto

11

Channels

4	100 Diameter clayware half round straight main channel 650 long bedded and jointed in cement mortar (1:1) (W.7.5)

2	100 Ditto 1150 long ditto

2	100 × 150 Diameter clayware half round tapered straight main channel 1250 long ditto (MH 4,8

1	150 × 225 Ditto 1250 long ditto (MH 9 & 225 Diameter clayware half round straight main channel 1250 long ditto (MH 11 & 225 Diameter half round curved main channel 1200 girth ditto (MH10

12

STANMORE STREET

NO 11 MANHOLES (cont)

Branch
bends

18	100 Diameter clayware three quarter section branch channel bend bedded and jointed in cement mortar (1:1) (W.7.5)

1	150 Ditto

Interceptor

1	225 Intercepting trap (Reference NAT 132) with 100 diameter cleaning arm including lever locking stopper with galvanized chain and staple and build trap into side of brick manhole and joint to pipe and bedding and surrounding in sulphate resisting cement concrete (14 N/mm² – 20 aggregate) (W.7.5)

13

Covers

11	Drainage castings (Bun Bros) medium heavy duty recessed manhole cover (No L 5255) with 750 x 600 clear opening with frame bedded and haunched in cement mortar (1:3) and cover filled with concrete (14 N/mm² - 20 aggregate) and sealed with grease and sand (W.7.5)

Step irons

MH 2	4
4	4
8	4
9	7
10	7
11	7
	33

33	Malleable cast iron step iron to BS 1247 Type A built into brickwork and make good fair face and flush pointing (W.7.5)

Item	Provide and hand to Architect a set of lifting keys to suit manhole covers

(End of No 11 Manholes)

14

199

SCHEDULE OF MAIN RUNS

Run	Length	Depth to invert	Depth to invert	Average	Item	Bed 100	Bed 150	Bed 225	Bench 100	Bench 150	Bench 225	Surround 100	Surround 150	Surround 225	Pipes 100	Pipes 150	Pipes 225	Next roadway	Ducts through foundation
MH1 – MH2	23 800	1100	1767	1434	1500	23.80						23.80			24.25				
MH3 – MH2	14 400	1100	1767	1434	1500	1440						14.40			14.85				
MH2 – MH4	16 200	1767	1815	1791	1450	16.20			4.20			12.00			16.15				2/100
MH4 – MH9	20 000 / 8 600	2000 / 2000	2429 / 1815	2215 / 1908	2000 / 2250		28.60			19.00			9.60			20.00			1/2
MH9 – MH6	26 000	1011	1464	1282	1250	26.00			26.00			26.45						26.00	
MH7 – MH6	11 200	1100	1464	1282	1250	11.20			11.20			11.40			11.65			11.20	
MH6 – MH8	16 400	1464	1817	1641	1750	16.40			5.00			11.40			16.85				2/100
MH8 – MH9	9 800 / 4 200	2000 / 2000	2429 / 1817	2215 / 1909	2250 / 2000		14.00			14.00		14.00				14.45			
MH9 – MH10	16 400	2429	2594	2512	2500			16.40			16.40		16.40				16.85		
MH10 – MH11	10 400	2594	2682	2638	2750			10.40						10.40			10.85		
MH11 – Sewer	By Local Authority					PC Sewer £ 1200													
Totals						108.00	42.60	26.80	46.40	33.00	16.40	76.60	26.00	10.40	110.50	43.45	27.70	37.20	

15

16

STANMORE STREET

MAIN RUNS

(See Schedule)

23.80	Excavate pipe trench
14.40	to receive pipes

not exceeding 200
diameter commencing
at reduced level
not exceeding 2.00m
maximum depth
and average 1500
deep including
earthwork support
to both faces, grading
and compacting
bottoms, backfilled
for a depth of 450
over pipe with
selected fill material
tamped by hand
in 150 layers and
with normal fill
material to surface
and remove surplus
excavated material
from site

(W. 3)

16.20	Excavate trench for
16.40	ditto but average

1750 deep

8.60	Ditto but not
4.20	exceeding 2.00m

maximum depth
and average
2000 deep

17

26.00	Excavate trench
11.20	for pipes not

exceeding 200
diameter
commencing at
reduced level
not exceeding
2.00 m maximum
depth and
average 1250 deep
as before – next
roadway

(W. 3. 6)

20.00	Ditto but not
9.80	exceeding 4.00 m

maximum depth
and average
2250 deep (not
next roadway)

16.40	Excavate trench

for 225 pipe
commencing at
reduced level but
not exceeding
4.00 m maximum
depth and
average 2500 deep
(not next roadway)

10.40	Ditto but average

2750 deep (not
next roadway)

Item	Keep the surface

of the site and
the excavations
free of surface
water

(D.25)

18

MAIN RUNS (cont)

Plain sulphate resisting cement concrete (14 N/mm² – 20 aggregate)

108·00	150 x 450 Wide bed laid to falls under 100 diameter pipe (W.5)
42·60	150 x 500 Ditto under 150 diameter pipe
26·80	150 x 575 Ditto under 225 diameter pipe
46·40	150 Wide benching to both sides of 100 diameter pipe (W.5)

19

Plain concrete (cont)

33·00	150 Wide benching to both sides of 150 diameter pipe
16·40	150 Ditto to 225 diameter pipe
75·60	150 Thick covering to 100 diameter pipe (W.5)
26·00	150 Ditto to 150 diameter pipe
10·40	150 Ditto to 225 diameter pipe

20

STANMORE STREET

MAIN RUNS   (cont)

		'Hepseal' vitrified clayware pipes and fittings with flexible joints laid in trench bottoms on concrete beds (measured separately) (W.6.1)		4	225 Diameter clayware duct 900 long and build into mass concrete foundations for and including packing around 100 diameter drain pipe with expanded polystyrene and including additional concrete, excavation and disposal below drain as necessary to give a minimum concrete base of 300 (W.6.6) (Blocks A & B
110·50		100 Diameter pipe			
43·45		150 Ditto		2	300 Ditto 900 long for and including packing around 150 diameter drain pipe ditto (Block C
27·70		225 Ditto			

SCHEDULE OF BRANCH DRAINS

No	Location	Length 100	Length 150	Depth of trench Min	Max	Average	Bends 100	Bends 150	Junctions 100	Junctions 150	Junctions 225	Builders' work Hole in wall	Hole in slab
1	VP to MH 1, 7, 10	12.00		450	900	675	6					3	3
2	G1,2,4,5,6,8,9 19,20,21,22,24,25	13.00		450	1100	775	26		13				
3	SVP to Block B	30.00		450	1100	775	12		6			6	6
4	SVP to Block A	30.00		450	1100	775	12		6			6	6
5	SVP to MH 3	5.00		450	900	675	2					1	1
6	SVP to MH 5	4.00		450	900	675	2					1	1
7	G3, 7, 19, 23	10.00		450	1100	775	8		4				
8	G17 to MH 5	3.50		450	900	675	2						
9	SVP to Block C	20.00	o	450	2429	1440	10			3	2	5	5
10	G28, 31, 32, 35, 38	5.00		450	2429	1440	10			2	3		
11	G29, 37	4.00		450	2429	1440	4			1	1		
12	SVP to Block D	19.00		450	2429	1440	6					3	3
13	SVP to MH 4	13.00		450	1615	1033	2					1	1
14	G42 to last	4.00		450	600	525	2						
15	G10 to MH 4	15.00		450	1615	1033	2		1				
16	G11 to last	1.50		450	1000	725	2						
16A	G12 to MH 4	6.00		450	1615	1033	2		1				
17	G13 to G14	2.50		450	600	525	2		1				
18	G14 to run	8.00		600	2112	1358	-			1(SD)			
19	G15 to last	4.00		450	700	575	2		1				
20	G16 to run	12.0		450	2112	1358	-			1			
21	G26 to MH 8	15.50		450	1617	1024	2		2				
22	G27 to run	4.50		450	1200	975	2		1				
23	G30/36 to run		4.60	1250	2429	1840	-	4		1(SD)	1(SD)		
24	G41 to run	14.00		450	2450	1450	2				1		
25	G33/34 to last	5.00		450	1450	950	4						
26	G39 to MH11		4.50	1250	2600	1925	-	2			1(SD)		
27	G40 to MH 11	9.00		450	2429	1440	2						
		270.00	8.50				126	6	35	8(100)	5(SD)	26	26

STANMORE STREET

BRANCHES

(See Schedule)

2/	4.00		
	2.50		

Excavate pipe (14 + 19
trench to (17
receive pipe
not exceeding
200 diameter
commencing at
reduced level
not exceeding
1.00 m maximum
depth and
average 500
deep as before
described

	12.00	
	13.00	
2/	30.00	
	5.00	
	4.00	
	10.00	
	3.50	
	1.50	
	5.60	
	1.40	
	1.10	
	5.30	
	6.10	
	7.00	
	7.30	
	2.10	
	3.90	
	1.90	
	2.60	
	2.80	
	3.30	
	2.50	

Ditto average (1
750 deep (2
(3+4
(5
6
(7
(8
(16
(9
(10
(11
(12
(13
(15
(21
(18
(24
(22
(25
(16A
(20
(27

25

6.90	
8.00	
8.20	
2.60	
2.40	
3.20	

Excavate pipe trench (13
to receive pipe (15
not exceeding
200 diameter (21
commencing at
reduced level (22
not exceeding
2.00 m maximum (25
depth and (16A
average 1250
deep as before
described

10.10	
2.50	
2.00	
9.60	
5.90	
9.20	

2/	2.50	
	4.50	
	7.10	

Ditto average (9
1500 deep (10
ditto (11
(12
(18
(20
(23+26
(27
(24

4.30	
1.10	
0.90	
4.10	
1.50	
2.00	
3.00	
2.00	

Ditto average (9
2250 deep (10
ditto not (11
exceeding
4.00 m (12
maximum depth (23
(26
(24
(27

26

205

BRANCHES (Cont)

270.00	150 x 450 Wide Concrete bed as before & 150 Wide benching to both sides of 100 pipe as before	
8.50	150 x 500 Wide Concrete bed as before & 150 Thick surround to 150 diameter pipe as before (under road	

27

Surrounds under
road & buildings

7/	2.25	150 Thick surround to 100 diameter pipe as before & Ddt 150 Wide benching to both sides of 100 pipe as before
	1.00	
	30.00	
	5.00	
2/	2.50	
	19.00	
	9.00	
	14.00	
	5.00	
2/	9.00	100 Diameter 'Hepseal' clayware pipe as before
	30.00	
	5.00	
3/	4.00	
	3.50	
	20.00	
	19.00	
	13.00	
	15.00	
	6.00	
	8.00	
	12.50	
	15.50	
	4.50	
	14.00	
	9.00	

28

STANMORE STREET

BRANCHES (cont)

				124		Extra over 100 pipe for bend (W.6.5)
2/	13.00 10.00 5.00 4.00 1.50 2.50		100 Diameter 'Hepseal' clayware pipe in runs not exceeding 3.00 m long laid in trench bottoms on concrete beds (measured separately) (In No 28 runs) (W.6.3)	6		Extra over 150 pipe for bend
				35		Extra over 100 pipe for single curved oblique branch junction 100 diameter
	4.50		150 Diameter 'Hepseal' clayware pipe as before described	8		Extra over 150 pipe for ditto
				1		Extra over 150 pipe for 150 diameter branch junction
	4.00		150 Ditto in runs not exceeding 3.00 m ditto (In No 2 runs)	7		Extra over 225 pipe for 100 diameter branch junction
				2		Ditto but 150 diameter branch junction

29

30

BRANCHES (cont)

26	Hole through 260 brick cavity wall in common bricks in sulphate cement mortar (1:3) for 225 clayware duct 300 long including expanded polystyrene packing between 100 diameter drain pipe and duct and make good brickwork
	&
	Hole through 150 reinforced concrete floor for 100 diameter drain pipe and make good

GULLIES

(See Schedule)

	Accessories jointed together and to drain with sulphate resisting cement mortar (1:2) and setting on and surrounding with plain sulphate resisting cement concrete (14 N/mm² - 20 aggregate) 150 thick including all necessary formwork, additional excavation, earthwork support, levelling and compacting and removal of surplus excavated material from site
13	150 × 150 Brownware (A accessible trap gully (Reference (B NAT 191) with 'S' trap and 100 diameter outlet and 100 diameter vertical back or side inlet with 'Stanford' joint stopper and 150 × 150 locking grating (Reference NAT 1009)
1	
	(W. 6. 6)

31

32

STANMORE STREET

GULLIES (cont)

10	150 × 150 Brownware (A gully (Reference NAT 191) as last but with 150 × 150 sealing plate (Reference NAT 1006) (B
5	(F
5	

1	150 × 150 Brownware (C non-accessible trap gully (Reference NAT 174) with 'P' trap and 100 diameter outlet, 100 diameter horizontal inlet and 150 × 150 sealing plate (Reference NAT 1006)

1	150 × 150 Brownware (D gully (Reference NAT 191) as before but with one 100 diameter and one 100 diameter vertical inlets, 'Stanford' joint stopper and 150 × 150 locking grating (Reference NAT 1009)

3	150 × 150 Brownware (E gully (Reference NAT 191) as last but with no inlets

3	400 Internal diameter brownware (G road gully with 150 diameter outlet and 'Stanford' joint stopper (Reference NAT 213) and 400 × 445 road gully grating (Drainage Castings, Burn Bros Reference DC 5791)

33

34

SEWER CONNECTION

Provide the Provisional Sum of £1200·00 for one sewer connection executed complete by the Local Authority

(W.8/B.11.1)

GENERALLY

| Item | Allow for providing assistance and apparatus for testing all drains in the presence of and to the satisfaction of the Local Authority and make good defects and re-test until passed (W.9) |

Item — Allow for providing assistance and apparatus for testing all drains in the presence of and to the satisfaction of the Local Authority and make good defects and re-test until passed

(W.9)

Item — Allow for flushing out all drains on completion of works

Item — Allow for protecting the whole of the work in this section

(W.10)

35

36

STANMORE STREET

To take

All soil and vent
pipes to have an
access plate at
ground floor level

Preamble notes

Drainage pipes and
fittings to BS 65 and
450

Workmanship and
materials to conform
with CP 301

37

Fresh air inlet
MH 11

| 1 | 100 Diameter cast iron fresh air inlet with hinged locking grille as manufactured by Drainage Castings (Burns Bros) catalogue reference DC 065 plugged and screwed to brickwork and jointing to and including 900 length of 100 diameter salt glazed stoneware pipe and one bend and including any necessary additional excavation and surrounding pipe in sulphate resisting cement concrete (14 N/mm² - 20 aggregate) and building in pipe to side of engineering brick manhole |

End of drainage example

38

211

APPENDIX

Computers and measurement

In this book we have shown how measurements should be taken from drawings and entered on to dimension paper. Current and future developments in computer technology will have an influence on this procedure and the likely changes are briefly discussed here.

There are three main probable areas of change:

1. The automatic generation of quantities at the same time as the drawings themselves are produced.
2. The entering of quantities by electronic means from the drawings.
3. The processing of quantities, which have been taken off manually, by computer.

Automatic generation of quantities

In a number of fields, such as the design of products, components, etc, a computer is used in the design. The designer works at a computer work station and sees his design develop on a visual display unit (screen). When the design is complete, the computer will not only produce the detailed drawings and calculations but also print out specification notes, bills of quantities, etc. This approach is widely used in many areas of engineering such as civil and structural engineering where designers follow recognized design procedures and mathematical formulae. Therefore, it is logical for the computer to calculate the quantities as well as calculating the stresses, loads, etc. of the item being designed.

In any field where quantities can be automatically generated then it is obvious that they should be so produced. The quantity surveyor's involvement will be in describing which items should be measured according to the *Standard Method of Measurement* and producing the standard descriptions for inclusion in the computer's memory. Paterson* describes the approach adopted at West Sussex County Council for building projects and is the best known description of the stages involved in the automatic generation of quantities for the type of projects measured in this book.

Automatic generation of quantities will only occur when either the design procedures are well defined and which can then be computerized or when the product is part of a standard system where a good data base, including quantities, exists.

The methods described in this book for describing finished items of work will therefore apply to automatically generated quantities. However, once the descriptions have been included in a computer file, they will not need to be written out again.

* *Paterson, J.W. (1977), Information Methods, John Wiley and Sons, London.*

Entering quantities by electronic means

In some cases drawings will have been produced where, although it was not possible for quantities to be automatically generated, the drawings have been so accurately drawn to scale that they can form an accurate basis for direct entering of measurements. As an item of work is identified as needing to be measured, the taker-off first enters on the computer typewriter keyboard the item reference number from a standard library bill of quantities. The drawing, from which the measurements are to be taken, is placed on an electronically sensitive drawing board. The dimensions of the items to be measured are entered by sliding a 'puck' over the drawing and pressing a button on the 'puck' at the appropriate point. The 'puck' (about $10 \times 6 \times 2$ cm thick) and which is attached to the computer by a wire, has a window with cross hairs in it. When the button on the 'puck' is pushed this enters the point under the cross hairs into the computer's memory. Thus if a linear item is to be measured, placing the puck at each end of the line will enable the length of that item to be calculated in the computer. It is obviously necessary for the drawing to be firmly fixed (usually by adhesive tape) to the electronic drawing board.

As an example of this process, assume that an area of floor covering is to be measured. The following stages would be adopted:

(a)   Identify the item on drawing.
(b)   Locate the item description in computer library.
(c)   Enter the reference number from the library onto the computer typewriter keyboard.
(d)   Place the cross hairs on the 'puck' over each corner of the floor area in turn, pressing the button at each point.
(e)   Enter the command on the typewriter that the measurements are now complete.
(f)   Proceed to next item to be measured.

A good example of this system is CATO (computer assisted taking-off) developed by Elstree Computing Limited. The quantity surveyor would need to be familiar with the approach to measuring described in this book and then regard a system such as CATO as a means of dispensing with the need for a pen and paper. The subsequent stages of working up and bill production are then automatic and reduce much of the time and effort spent in a manual approach. It is also possible to enter dimensions via the typewriter keyboard if the drawing is not to scale or items which need to be measured are not shown on the drawings.

Computer processing of quantities

Quantity surveyors have used computers for processing quantities for many years. In the late 1950s and early 1960s systems were developed using large and expensive main-frame computers. Taking-off was largely done in a conventional manner and the descriptions were then coded by reference to a standard library of descriptions. In 1967 the RICS published a report on Computer Techniques which described 10 different systems for producing bills of quantities in this way. In 1973 when the report was updated there were more than 20 systems in everyday use.

In the mid-1970s there seemed to be a decline in the use of such systems but by the end of the decade microtechnology was giving an added impetus to the development of this type of approach. The increasing cost of labour for processing bills, the demand from clients for a quicker service and the much reduced cost of computers have all added to this drive. The recent publication by the RICS in 1981, *The Chartered Quantity Surveyor and the Microcomputer,* gives a very good review of the developments in this field. It is almost certain that developments in this area will continue at a rapid pace. However it will still require the surveyor to identify those items which need to be measured and to select them from whichever standard library of descriptions is being used.

Conclusions

In order to provide a good service to clients quantity surveyors will have to exploit computer technology to the full. For a description of developments which are on the immediate horizon, Chapter 4 of *Cost Planning and Computers,* published by the PSA Directorate of Quantity Surveying in 1981, is essential reading. In this report, which describes research undertaken at the Department of Construction Management of the University of Reading, it is quite clearly shown that an effective data base is essential for effective cost planning. Effective cost planning is a prerequisite for the procurement of efficient and timely building projects. To achieve effective cost planning computer technology must not only be exploited at all stages of the quantity surveying process but also all those stages must be integrated.

Quantity surveyors must therefore carefully examine their procedures and see the extent to which their present skills can be enhanced by the application of computer technology. It is a path which will require both financial and intellectual commitment and is not without pitfalls. However, it is a path which must be taken.

Index

The order of this index is not alphabetical but follows the layout of SMM6. Where an item appears more than once in any of the eight examples of measurement, the page number refers to the first appearance of that item in each example concerned.

216